Writing Spaces: Writing as Transformative, Scholarly and Creative Practice

At the Interface/Probing the Boundaries

VOLUME 119

The titles published in this series are listed at *brill.com/aipb*

Writing Spaces

Writing as Transformative, Scholarly and Creative Practice

Edited by

Esthir Lemi
Ekaterina Midgette
Jessica Seymour

BRILL
RODOPI

LEIDEN | BOSTON

Cover illustration: *Studying at the Library*, Esthir Lemi, 2000 (oilpainting 1.00x 1.50 m)

The Library of Congress Cataloging-in-Publication Data is available online at http://catalog.loc.gov

Typeface for the Latin, Greek, and Cyrillic scripts: "Brill". See and download: brill.com/brill-typeface.

ISSN 1570-7113
ISBN 978-90-04-39430-8 (paperback)
ISBN 978-90-04-39431-5 (e-book)

This book is printed on acid-free paper and produced in a sustainable manner.

Contents

Notes on Contributors

David W. Bulla
is Professor of Communication at Augusta University in the United States. He is the author of several books about Abraham Lincoln and the press, including *Lincoln Mediated* (Transaction, 2015), which he co-wrote with Gregory A. Borchard.

Nathan James Crane
is an interior architect, design thinker, writer and educator. His current research focuses on the inherent spatiality of writing practices, particularly those of French Poststructuralist critical theory. Nathan is currently design writer and editor for The Adelaide Review and a doctoral candidate, tutor and lecturer in History and Theory at the University of South Australia.

Phil Fitzsimmons
is currently Assistant Dean of Research in the Faculty of Education, Business and Science at Avondale College of Higher Education, Cooranbong, Australia.

Gail Hammill
is an Assistant Professor of English at the American University in Dubai where she teaches literature, composition and world culture courses. Gail is also a graduate of the School of American Ballet. She is currently working on a collection of personal essays exploring spirituality and embodiment.

Genevieve Jorolan-Quintero
is a faculty member of the Humanities Department of the University of the Philippines, Davao City. Her work is inspired by the research she does with the indigenous communities in Mindanao that aim to preserve the oral traditions. Her researches have enabled her to explore and write about the experiences and issues that confront these indigenous communities. She continues to share her research outputs to international conferences on culture and the humanities, and continues to draw inspiration from these researches for the stories she writes.

Syeda Hajirah Junaid
is a freelance researcher in Pakistan. Her background is in marketing and has worked in a Telecom company for 8 years. Her passion for writing, and her

responsibility to shed light on her ancestor's contribution to Urdu literature in the sub-continent and Pakistan, urged her to write this research paper.

Edie Lanphar
is currently Principal of Holy Cross Catholic School, in Ventura California.

Esthir Lemi
is a composer and a visual artist. She lectures as visiting researcher "Art and Anthropology" at the School of Fine Arts, Athens, Greece.

Imogen Lesser Woods
has recently completed her PhD at the University of Kent in Canterbury. Her background in architecture and a love of literature has led to her current research in the use of language in the architectural field.

Panagiota Mavridou
is an architect (NTUA, Athens, 2013) with an MA in Theory and History of Architecture (UPC, Barcelona, 2015). Wishing to understand how we inhabit, perceive and represent the world, she explores the relationships between body, language, space and narrative.

Sam Meekings
is an Assistant Professor of Writing at Northwestern University in Qatar. He is also the author of *Under Fishbone Clouds* (called 'a poetic evocation of the country and its people' by the New York Times) and *The Book of Crows*. His latest novel, *The Afterlives of Dr Gachet*, was published by Eyewear Press in summer 2017. His website is www.sammeekings.com.

Barış Mete
is an Assistant Professor of English currently teaching literary theory and the twentieth-century British novel at Selçuk University in Turkey.

Ekaterina Midgette
is Associate Professor of Literacy Education at the College of Saint Rose, Albany, NY, USA.

Sevil Nakisli
MA in Economics, MS in Education, is a teacher of Mathematics in Texas, USA.

Layla M. Roesler
teaches at the École Normale Supérieure de Lyon, in France. Her research subjects include the work of Yves Bonnefoy, literature and philosophy, the theory of the essay, the theory of difficulty.

Yadigar Sanli
is a freelance researcher and writer. She has an MA in 19th century Turkish Literature in Translations, and has a PhD in Contemporary Turkish Literature. She writes about culture, identities, minorities, language and literature. She currently resides in London.

Shelley Smith's
background is in academic and practice-based design, architecture and urbanism. The focus of current research revolves around experimentation with the written word as the conveyor of spatial experience and sensibility, and the development of alternative methods with which to present this.

Images

Introduction

Esthir Lemi, Ekaterina Midgette and Jessie Seymour

This book is a collection of the peer-reviewed papers presented at the fourth Inter-Disciplinary.net group Global Conference, Writing as Spaces. The conference brought from all walks of life and corners of the world, and the discussions were engaged in writing as a method, as a practice, and as a way of life. We are excited to share this collection with readers who are interested in space as it affects writing and writers.

In Part 1, 'Meeting the Student and the Self,' the reader is invited to look deeper into new and traditional forms of writing and their implications for teaching and pedagogy. Writing, be that in the form of an Instagram narrative, a personal reflection, or a story inspired by myths and folktales, is a powerful force that unleashes the creativity of both the author and the reader (regardless of age) and has the potential to originate new forms. The common theme of this section is helping students to find their writing space, discovering the boundaries of individual writing spaces, and ways to transcend them. Sam Meekings, an Assistant Professor of Writing at Northwestern University in Qatar and the author of several novels, explores how writing about physical spaces uncovers the intricate and mutually-dependent relationship between place and memory. Looking at the practices of writers such as W.G. Sebald and Guy Debord as well as the author's own writing process, Meekings suggests that writing about specific locations is often a way of conjuring ghosts, since place frequently functions in texts as a palimpsest in which different layers of time co-exist.

In contrast, Genevieve Jorolan-Quintero, a researcher, writer and faculty member at the University of Philippines, takes the reader to the vanishing world of the oral tradition and experiences of the indigenous people of Southern Mindanao. The goal of her research is to preserve and cherish the culture and wisdom of these peoples for generations to come. However, in the process of documenting ethno-epics, folktales, and myths, Jorolan-Quintero discovered her own writing space in the form of her own stories inspired by the experiences and the legends of the indigenous community.

Phil Fitzsimmons, an assistant dean at Avondale College in Australia, and Edie Lanphar, a school principal in the United States, take the reader to the world of a middle school classroom to show the effectiveness of the reading-writing connection in helping students to create a personal relationship with the text through writing. Using a young adult fantasy novel as a basis for reflective discussions and personal writing, the researchers helped their students uncover deeper memories and comprehend the spirituality found in their familial, emotional, and religious connections.

Writing as a spiritual space is the main theme of Gail Hammill' s paper. Hammill teaches literature and composition at the American University in Dubai, uses this paper to discuss the discrepancy between the ability of many writing teachers to inspire their students to create, discover, and reflect through writing, and their own lack of discipline to write. Hammill finds a solution to this dilemma by connecting with her inner spiritually and the spirituality that she found in the writing community of women who bring support and inspiration to fellow writers.

In the final chapter of this section, Ekaterina Midgette, who teaches literacy at the College of Saint Rose in the United States, and Sevil Nakisli, a graduate student of education at the same institution, explore another way to grow writers and connect with the intended audience: through humour. Midgette and Nakisli found that children and adolescents can use unprompted humour to make their persuasive appeal stronger and more engaging. However, humour appears to be underutilized in formal writing instruction despite its well-known potential to capture the recipient in the writing space of persuasion.

In Part 2, 'Meeting the Master,' David W. Bulla, a professor of communication at Augusta University in the United States, opens the section with an exploration of Mohandas K. Gandhi' s life as a writer. Bulla goes deep into Gandhi' s physical space as he was writing, as well as what and when he wrote, how he saw his writing, and what it meant to him on a spiritual and practical level. Gandhi stated in his autobiography that his responsibility as a journalist was to interact with and serve his readers, so Bulla' s interest in Gandhi' s life as a writer is particularly poignant in that regard.

Following on from Bulla, Syeda Hajirah Junaid examines the past and current challenges faced by three generation of writers. Junaid is a freelance researcher with a background in Marketing, and her ancestor' s contribution to Urdu literature in Pakistan inspired her to write this paper. Junaid' s chapter explores the different spaces encountered by the three women writers – her great-grandmother, – her mother, and herself – and how these spaces directly or indirectly impacted their output.

Barış Mete, an Assistant Professor of English at Selçuk University in Turkey, reminds the reader of the classical theory that lies behind the writings of an Anglo-Irish novelist-philosopher, Iris Murdoch. His research investigates how the Platonist conception of art as imitation has become in her fictional and philosophical writings the paramount issue of discussion.

In the fourth paper for this section, Layla Roesler examines the recurring trope of the 'vrai lieu' (or 'true place') Yves Bonnefoy' s poetry. Roesler, a teacher at at the École Normale Supérieure de Lyon, explores how Bonnefoy' s writing allows him to create a communion with the physical – essentially allowing him to create a physical space which is inhabited by symbols, and establishes an imaginary space of the 'here.' This 'here' is supported by the reader' s own attention. Roesler' s work demonstrates how Bonnefoy' s poetry builds on tools of difference and deixis to spatialize his writing.

To conclude the section, Yadigar Sanli, a freelance researcher and writer based in London, examines the space that Bediuzzaman Said Nursi (1878–1960) occupies amongst Islamic scholars in Turkish history. Said Nursi' s writings span 6000 pages, much of which was produced during an exile where he was kept under observation and stripped of many resources that we contemporary writers take for granted. The limitations of Said Nursi' s physical space did not prevent him from writing his philosophical interpretations of religious texts – his works straddling the line between academic works and fiction. Sanli explores Said Nursi' s physical context to draw conclusions about how space-time-material is important to a writer.

In Part 3, 'Meeting the Artist,' five chapters unveil patterns on which writing as an artistic act may be "a rightful heir to the scientist and her predecessor" (Nikos Karouzos 1990). Victor Hugo had foreseen, at the broad use of printing press and moveable type, a change in approaching art. We may take for granted the writing and printing process and through digital age, contemporary phenomenological approaches to writing have changed. The introduction of a new generation of scientists-artists may also influence the academic writing. The first two chapters of this section have been written by architects: Panagiota Mavridou and Shelley Smith. Mavridou presents 'The Movement-Spatiotextual Inscriptions,' where she uses walking as a tool for creative investigation both on writing and spaces. Mavridou's paper presents three stages of an open-ended experiment designed to explore, register, expose, and comprehend the creative process of writing as it relates to space.

Smith's 'At the Crossroads: Writing Spaces between Academia and Embodiment', meanwhile, analyses how travels manifest the "commitment of an embodied and sensual response"; a transition between the space of academic writing and an "expressive, sensorial practice." As academic writing has its strict rules in style, Smith explores the space of the crossroads of her own writing practice by creating a dialogue between the sensorial, perceptual voice of non-representational inquiry – a more poetic, embodied practice – and the reflexive, cognitive voice of academic writing – a cartographic position.

Esthir Lemi' s paper explains the 'Artist as a Writer' condition, from the composer and visual artist' s perspective by linking analogue and digital technology. Lemi argues that in order to preserve visual material, some text needs to be written. She proposes her writing pattern derived from musical composition, which is finalised in the form of a book, presenting her method as a common ground of all arts and the realisation of these ètudes.

In the fourth paper for the section, architect and researcher Nathan James Crane suggests an "expanded spatial practice." He writes that this, in effect, constructs and cultivates "space for thought" through the process of writing. His paper, 'Exploring the Written Wor(l)d: Writing as a Spatial Practice,' develops and elaborates on the concept of writing as spatial practice with particular reference to French Postructuralist writers such as Hélène Cixous and Jacques Derrida.

Finally, Imogen Lesser Woods introduces 'The Literary Spaces of Mervyn Peake's The Gormenghast Trilogy Used as a Foundation for Architectural Exploration.' Mervyn Peake, an artist, illustrator, poet and writer, produces an imaginative space that we as readers are able to approach. This ability to inhabit literary spaces, not through physical acts but through the potential of the architecture of a narrative, is one of the benefits of the common space of communication that is created by the artistic act of writing, and the pleasure of reading as well.

PART 1

Meeting the Student and the Self

CHAPTER 1

Writing Ghostly Spaces: Place as Palimpsest

Sam Meekings

Abstract

This chapter will explore the relationship between place and memory through writing. It will examine W.G. Sebald's palimpsest landscapes, the roles of psychogeography and dérive in exploring the past, and how the idea of writing about place is bound to the idea of ghosts. It will also suggest some of the ways in which spatial travel is analogous to time travel in life writing. I will argue that any form of movement in space can, and frequently does, inspire and facilitate a corresponding movement through time for the writer seeking to document, reveal, uncover and remember the past. Furthermore, the chapter will demonstrate how this movement might lead to a reconsideration of the relationship between place and time in writing.

Keywords

Sebald – place – dérive – Derrida – memoir – travel

•••

Returning to a familiar place awakens memory. The journey back to a childhood home, or a city one used to live in, stimulates the senses and so sparks flashback, reminiscence, nostalgia. Through travel, then, we awaken ghosts. Indeed, place has a complex relationship with recollection, in particular as a trigger. In a number of literary texts, place functions as a palimpsest in which different 'layers' of time co-existed and might, therefore, be accessed at any particular moment. In other words, certain places are haunted: they are employed in writing as vessels which contain and reveal the experiences that happened therein. Therefore, the writer must be aware that by writing about place, one is also always writing about time. Writing about place often means writing about ghosts.

Consider W.G. Sebald's *The Rings of Saturn*, a travel narrative in which a walking tour of the coastline of East Anglia becomes a profound meditation on loss, ruin and destruction. In Sebald's work, each location leads to discussion

 | DOI:10.1163/9789004394315_002

of fragments of the past. Speaking of a house in Suffolk, for instance, the narrator remarks upon how difficult it is to 'say what decade or century it is, for many ages are superimposed here and coexist'.[1] This conception of place as a palimpsest in which different pasts – both public and private – all overlap suggests that memory is not only internal but also external. It is dependent upon physical spaces. Place thus exists for Sebald as a kind of archive, a place where memory is reawakened and revealed.

This mingling of the past and present in Sebald's collecting together of events and items of folklore serve to construct what Gregory-Guilder calls Sebald's 'collated, collaged landscapes', where the past is as alive as the present in each location.[2] Indeed, each place visited functions as a tapestry of connections. Gregory-Guilder goes on to point out that 'Each place in Sebald's work thus opens onto a ghostly version of its past existence, allowing a kind of traffic between the present and the past, the living and the dead'.[3] Place is 'ghostly' in Sebald's text precisely because it transcends a linear chronology as a vessel of memories and histories. For example, visiting Southwold prompts Sebald to recall reading about a naval battle with the Dutch that occurred off the coast there in 1672, while a trip to Dunwich leads him to reflect on how this village was once 'a place of pilgrimage for melancholy poets of the Victorian age'.[4] For Sebald, locations are not merely geographic, composed of landscapes, topographical features, architecture, and terrain, but are instead composed of the events and experiences that occurred within them. Place is then, in some senses, reliant on memory, and as such is a product of composition.

To write about a place is to play a role in the creation and continuing existence of that place. Our sense and conception of different spaces is dependant on our experiences in them. As Simon Schama has suggested, place is constructed as much by imagination as by facts: 'Before it can ever be a repose for the senses', writes Schama, 'landscape is the work of the mind'.[5] In other words, locations change depending on who is viewing them. The pyschogeographic traffic runs two ways: the emotions and behaviour of the viewer are influenced by the location but the location is also changed by the emotions and behaviour of the viewer. In much the same way, it might be suggested that the landscape of a text is constructed by the narrative, while at the same time the narrative is a product of that landscape.

---

1 W.G.Sebald, *The Rings of Saturn* (London: The Harvill Press, 1998), 36.

2 Christopher C. Gregory-Guider, The 'Sixth Emigrant': Traveling Places in the Works of W.G. Sebald, *Contemporary Literature* 45.3 (2005): 428

3 Ibid.

4 W.G.Sebald, *The Rings of Saturn* (London: The Harvill Press, 1998), 49.

5 Simon Schama, *Landscape and Memory* (London: Harper Perennial, 2004) 6–7.

This is important for me. On a hot day in early summer, my brother pulled his car onto the side of the road on his way home and passed out. He never woke up. He was twenty-four. It was in the aftermath of this event that I decided to write about him. I planned to record his life, by collecting and writing down a history of our time together. I would write about our shared past, then, in order to attempt to 'save' something of my brother. It seemed obvious to me that to do so I would need to retrace our steps: to revisit the houses in which we had been born, grew up, spent our childhoods; to see again the schools in which we had studied; the cities in which we had made our homes. In this way, I might unearth the memories that would sustain a narrative.

I was, therefore, drawn to the Situationists' idea of the *dérive* (or drifting) as a key source of inspiration. The idea of the *dérive* is that getting lost defamiliarises the world, and so is itself an almost creative act. The world becomes momentarily strange in the moment of getting lost. Guy Debord, writing about dérive from a Situationist perspective, talks specifically about 'the sudden change of ambience in a street within the space of a few metres...[and] the path of least resistance which is automatically followed in aimless strolls'.[6] Drifting thus functions almost as a cousin to automatic writing, whereby the subconscious mind dictates the direction and content of the ensuing text. Once again, we are led back to Sebald's conception of place as a palimpsest which might, at any moment, conceal or reveal something to the visitor.

It might be added that wandering in physical spaces has always been a central tool of historical inquiry. Richard Cobb, for example, has argued that there is no better way to access the past than to find it by exploring the specific physical locations of the present: 'A great deal of Paris eighteenth-century history', he writes, 'can be walked, seen, and above all heard, in small restaurants, on the platform at the back of a bus, in cafes, or on the park bench'.[7] A place is a composite of all the historical events and ages that belong to it, and so by travelling to different spaces (and indeed by writing about them) we are accessing different elements of that composite. By exploring a place, we uncover not only its present state but also its history.

However, when these lines between past and present are blurred, just as in Sebald's 'collated collages', a narrative becomes open to the possibility of the uncanny. After all, when we say that certain places are 'haunted', what we mean in essence is that they retain the imprint of events that occurred there. The role of 'ghosts' in many fictional works depends upon the idea that they are tied to place while simultaneously being dislocated in time. One result of location dictating chronology, therefore, might be a conjuring of ghosts. According to

6 Guy Debord, 'Theory and Derive', *Les Levres Nues* #9, (November 1956):11.

7 Richard Cobb, *A Second Identity* (Oxford: Oxford University Press, 1956):19–20.

Jacques Derrida, there is no escape from the ghostly nature of the world, since 'everyone reads, acts, writes with his or her ghosts';[8] he argues, indeed, that we are 'structured by the phantasmic and in particular that we have a phantasmic relation to the other, and the phantamicity of this relation cannot be reduced'.[9] We are all, he implies, inhabited by ghosts, and it is arguably the case that any work which is concerned with excavating and unearthing the past cannot escape the notion of the ghost at the shoulder, and indeed of the ghost within.

Ghosts, of course, complicate any simple sense of space as solely physical or geographic. As Roger Lockhurst remarks, 'Only ghosts, after all, can walk through walls, [or] breach the boundaries of the increasingly privatised zones of the city'.[10] This mobility is one that finds its mirror in narrative, for in the same way that ghosts can pass through walls, so a story can transcend the physical and chronological limitations of its starting point. Any writing about place also suggests a breaching of boundaries, a blurring between the public (the physical record of the text itself) and the private (the personal experience). Much writing about place might thus be said to be a way of generating, mediating or voicing ghosts.

My own search for my brother's ghost illustrates something of this mobility and the anxieties that it produces. In order to excavate our shared personal history, I went looking for ghosts. I headed, naturally enough, to my grandparents' farm:

In the early hours of open roads it does not take me much more than an hour to reach the farm. We never thought to call it by its full name. 'The farm' says it all. This is the rundown corner of East Sussex where Mum grew up. I park beside the shop. The roof is battered and rusty, the white paintwork faded, and the front windows cloudy with dust and cobwebs. Lined outside the door are several bags of compost, one of the few things now sold here, and which my grandfather still loads into the boot of his car before delivering each week to neighbouring farms and allotments. Tucked in amidst the sagging plastic is an I.O.U. scrawled on a torn piece of notepaper.

My grandfather must have seen the car from the window, as he is now waiting for me at the door. He does not ask why I have come. Nor does he remark on how early it is. He gestures for me to come inside. In the days when my brother and I had chased each other around our grandparent's ramshackle farm, the shop was usually filled with locals looking over the many boxes of fresh

8 Jacques Derrida, *Spectres of Marx* (Oxford: Psychology Press, 1969): 19–20.

9 Jacques Derrida and Maurizio Ferraris, *A Taste for the Secret*, in Gray Kochhar-Lindgren's *TechnoLogics: Ghosts, the Incalculable, and the Suspension of Animation* (New York: State University of New York Press, 2005):199.

10 Roger Lockhurst 'Occult London', *London: From Punk to Blair* (London: Reaktion Books, 2003):337.

mushrooms or else picking up punnets of strawberries and potted plants. Now the room is filled with countless empty polystyrene boxes stacked haphazardly against the walls.

My grandfather's office is the tiniest of cubbyholes harbouring a wooden chair, a small safe, a calculator, a pad of post-it notes and a whole heap of inkless biros. He hands me a cardboard box full of old paperbacks he has found in one of the packing sheds. *Have a look. See if there's anything that takes your fancy.* It is impossible to say how long the paperbacks might have been hidden away back there, though a few at the top of the pile have clearly been shrivelled and curled by years of damp. It always amazes me that, though most of my grandfather's sheds and greenhouses fell into disuse a long time ago, he is forever rediscovering all manner of things. Boxes left in one shed might vanish and be given up for lost only to turn up a decade later in another of the outhouses, under an upturned lawnmower, or a reel of fencing. When I was young the back half of the farm had seemed a strange and mythical place where time did not operate as it did elsewhere, and my brother and I often used to imagine that each shed might provide a gateway to another age – one long in the past or far off in the future. My grandfather's latest finding brings my childhood imagining back to me.

Is it alright if I take a look around? My grandfather nods. *Of course. Looking for anything in particular?* I hover in the doorway. *I'm not sure.* I leave him to his accounts, turning from the tiny office towards the storeroom at the back of the building. I make my way through a maze of broken crates and a heap of ancient newspapers. I open one at random and am disappointed to find that the headlines, local stories and little histories have all bled together into a dark and illegible blur.

Outside the door at the back end of the shop is a large, rusty vat that has filled up with rain. I make my way around it and follow a small alley between several rows of breeze-block mushroom houses that remind me of a stretch of bunkers set out for some as-yet undefined emergency. A few feet across from them, the packing sheds are crammed with all manner of useless odds and ends. I am struck again by the same feeling of uncertainty and expectation that I used to feel when beginning a game of hide and seek. When it was my turn to be the seeker, I would stand beneath the crocodile and count to one hundred before bursting through the back door and starting down that alley, knowing that my brother might be holed up in any of the buildings before me, hidden amid the old gardening tools, rusty deckchairs, and stacks of cardboard boxes growing moss. Perhaps that boisterous child is still there, curled up behind an old tractor tyre or mouldy sofa, torn as ever between the urge to stay still and silent and so prolong the game and the equally strong craving to leap up suddenly and start screaming.

I carry on into the alley between the buildings. I start down the right-hand side of the line, pressing my hands against each of the mushroom bunkers in turn. Each one is padlocked and will not budge. They had always been impenetrable when we were young and, now that I think about it, I realise I have never seen the inside of a single one of those whitewashed buildings.

We were told they were filled with mushrooms and we accepted this as the truth without ever asking for proof or verification. However, our grandfather could have told us that he was growing miniature orang-utans and baboons in there and we probably would have believed him, for he spoke with such casual authority. Besides, as little children we took it for granted that adults would never lie to us.

I push against each padlocked door in turn, hoping that one of them might open just a crack to let me catch a glimpse of whatever is kept inside. None of them will yield. The reason we were given for these always being kept locked was that mushrooms grow only in pitch black, taking root from shadow. This, by the way, is almost precisely what some of the early church fathers said of the soul: that it is added mysteriously in the darkness of the womb. Perhaps that might explain why so many children are afraid of the dark. I picture a child locked inside one of those sheds, and feel a shudder drip down every knot of my spine. I pound against the heavy door with my fists, rapping out our secret code – three knocks, pause, then two, pause, then another three. Then I press my ear hard up against it, but all I can hear is my heart pounding out its own secret code.

I turn quickly from the bunkers and hurry over to the sheds on the opposite side. I am jogging now, certain that time is running out, though where it is going I cannot say. I explore shed after shed, pushing through the thick cobwebs that hang down from every corner. I find old bags of fertiliser, wooden boxes, coils of hose, an assortment of troughs, spare parts for the cement mixer, wobbly chairs chewed by woodworm, a few flat tyres and an endless supply of polystyrene crates. I grow more frantic with each new shed I enter. I am soon in a mad rush, shoving aside great piles of bubble-wrap and knots of plugs and wires. I rifle through old boxes, splintering cupboards and chests of drawers, nudging aside with my shoes any bags or cartons I do not dare disturb for fear of the insects or reptiles I might find making their homes within.

I find nothing. I leave the final shed and turn out of the alley toward the small stream that marks the end of the farm. I am out of breath, and need to sit down. Beside the slow trickle of water stand the bones of a tractor, its frame turned a turgid brown where the bright orange paint has curled clean from the rusty skeleton. Tyres, seat and gearbox have all been stripped away, and the rest has been put out to pasture beneath a few stunted apple trees. The tractor is so frail that it is hard to believe it can still stand unaided. It puts me in mind of those reconstructions of dinosaurs that fill the main halls of museums of natural history, their giant bones suspended by the finest threads of translucent string. My brother and I climbed on it many times, appropriating it as a getaway vehicle or makeshift time machine, but I have no doubt that if I even brush it with my fingers now it will collapse to a heap of dust. That we had never seen it move did not bother us, for like everything else there we felt it belonged more to our fantasy world of make-believe than to the real world that existed beyond the farm's borders. Like many of the curios we came across in the packing sheds, its own history had finished long ago and it served to remind us that things had been different before we were born, and would be different too when we grew older.

There is nothing for it. I have to go back. I turn and start towards the shop and my grandfather. I have kept him waiting long enough, and know there is no way I will be able to explain to him what I have been looking for. At first I am disappointed that my ramble around the farm has yielded so little – no hidden secrets, no hints, no crocodiles; but soon I start to feel oddly upbeat. I am actually relieved that so many of the doors are locked, for had I been able to look inside and then found nothing I would have felt crushed. This way I can at least hold on to the possibility, however slim, that within the crumbling buildings there is something of vital importance. Something.

My own wanderings in search of my brother therefore suggest that place and time are intimately connected, and cannot be separated from one another. The past is bound to its ghostly spaces.

Perhaps then memory can never be fully disentangled from its location, regardless of the number of places visited and past experiences re-awakened. The boundaries between place and memory are constantly being blurred. This is not wholly surprising: all writing in some way reflects the writer, and frequently therefore the self of the author and the space that is written about by the author become conflated. Sprinkler acutely notes that 'No autobiography can take place except within the boundaries of a writing where concepts of subject, self, and author collapse into the act of producing a text'.[11] It is precisely this collapse that is demonstrated by Sebald having past and present, place and self, all come to overlap. And it is this collapse too that occurs in my own writing, where each location leads to a blurring between the experience of the present and the memory of what occurred in each space in the past.

In conclusion, the practice of writing is one of forging connections, particularly between the external and the internal. If writing about particular physical

11 Michael Sprinkler, 'Fictions of the Self: the End of Autobiography' *Autobiography: Essays Theoretical and Critical* (Princeton: Princeton University Press, 1980): 342

spaces summons ghosts, it is because our memories cannot be separated from the places in which they are formed and the people with whom we interacted there. Situating a text within a specific location often reveals the intricate and mutually-dependent relationship between place and memory. Whether in Sebald's conception of place as a palimpsest which both reveals and conceals, or in the dérive of Debord and the Situationists where the act of moving through new spaces triggers creativity, place is often ghostly in writing because the act of describing location also frequently functions as a way of describing our experiences, memories and selves within that location. It is in this way that, as Derrida asserts, we read, write and act with our own ghosts. In my own practice, writing about specific spaces allowed me to depict a blurring of time periods in order to emphasize the persistence and predominance of memory. In the end, it is perhaps ironic that writing about 'ghostly' spaces proves that the act and process of writing brings the past back to life.

Bibliography

Cobb, Richard. *A Second Identity.* Oxford: Oxford University Press, 1969.

de Man, Paul. 'Autobiography as De-Facement'. *Comparative Literature* 94.5 (1979)

Debord, Guy. 'Theory of the Derive' Trans. Ken Knabb. *Les Levres Nues* #9 (November, 1956): http://www.cddc.vt.edu/sionline/si/theory.html

Debord, Guy. 'Introduction to a Critique of Urban Geography' Trans. Ken Knab. *Les Levres Nues #6* (September, 1955) http://www.cddc.vt.edu/sionline/presitu/geography.html

Derrida, Jacques. *Spectres of Maoryrx.* Oxford: Psychology, Press. 1994.

Derrida, Jacques and Maurizio Ferraris. *A Taste for the Secret.* Quoted in Gray Kochhar-Lindgren's *TechnoLogics:Ghosts, the Incalculale, and the Suspension of Animation.* New York: State University of New York Press, 2005.

Gregory-Guider, Christopher C. 'The "Sixth Emigrant"; Traveling Places in the Works of W.G.Sebal'. *Contemporary Literature* 46.3 (2005).

Lockhurst, Roger. 'Occult London'. *London: From Punk to Blair.* Ed. Joe Kerr and Andrew Gibson. London: Reaktion Books, 2003.

Proust, Marcel. *In Search of Lost Time Vol1: Swann's Way.* Trans. C.K. Scott Moncrieff, D.J. Enright and Terence Kilmartin. London: Vintage Classics, 1996.

Schama, Simon. *Landscape and Memory.* London: Harper Perennial, 2004.

Seald, W.G. *The Rings of Saturn.* Trans. Michael Hulse. London: The Harvill Press, 1998.

Sprinkler, Michael. 'Fictions of the Self: the End of Autobiography'. *Autobiography: Essays Theoretical and Critical,* Ed. James Olney. Princeton: Princeton University Press, 1980.

CHAPTER 2

Finding My Writing Space: from Research to Storytelling

Genevieve Jorolan-Quintero

Abstract

My chapter explores the connection between research and creative writing- how one is essential to the other, and how one becomes the foundation for the other. Recently, I have put together an anthology containing ten (10) stories. Each story is inspired and based on my interactions with the indigenous people (IP) in Southern Mindanao in the Philippines. Two of the stories which were recently published in literary journals, – *Loom of Dreams* and *Invocation* – tackle the IP's struggle to preserve their culture and their notion of spirituality juxtaposed with the Christian religion. *Waterfall* brings out their concern for their environment, their ancestral domain, as it continuously faces destruction against the persistent lure of economic progress. In the process of writing these stories, I began to realize the crucial role of the writer, whether as researcher or storyteller, as the voice and the vehicle through which the experiences of others are brought to light. At first, my research goal was to salvage what remains of the oral traditions. There is urgency in this effort because the bards who can still chant and sing the narratives are getting few. As I recorded the ethno-epics, the legends, myths, and folktales, I learned more from the experiences of the indigenous people. So, I started to write my stories. In doing so, I found my writing space. My research work evolved into something beyond the technical, beyond recording of data, beyond reporting the empirical. I became drawn to the places, especially to the people in these places. And it is in these places, with the people, where I found my writing space. Every place I came to, every community I interacted with has provided a space for me. Every place has become a space for me to explore, to fill, to write about.

Keywords

Indigenous communities – indigenous people (IP) – Southern Mindanao – oral traditions – ethno-epics – legends – myths – folktales – bards – Philippines

•••

 | DOI:10.1163/9789004394315_003

When I learned about the Writing as Spaces conference, I had second thoughts about applying. I assumed it was only for writers, and I didn't consider myself as one. But I ended up writing and submitting an abstract anyway, and got qualified to present a paper.

I divided my chapter into four sub-topics: from fascination to research, from research to illumination, from illumination to imagination, and finding my writing space. This is an introspective chapter, a reflection on how anyone can find reason to write, motivation to write, a niche, a space.

1 From Fascination to Research

Tales of old, stories of magic and the supernatural, of deities and spirits of nature – these filled hours of my childhood both at home and at school. My favourite book aisle in the library was the one where fantasy books were shelved. That aisle became my reading space. At home, I would constantly rummage through my Mother's mini library in search of books on Philippine myths and legends, most of which came with colourful illustrations and pictures. This fascination lasted until 4th grade when reading mystery and suspense detective novels became the trend. Or so I thought.

Years later, at the university, I had to take a course on Philippine Folk Literature as a requirement for my degree. I was then re-acquainted with my childhood fascination. But this time, it was no longer reading but an in-depth study of literary texts: from myths, legends, and folktales, to ethno-epics. My childhood interest in folk literature made it easier to appreciate academic texts. This single course on Philippine Folk Literature became the springboard to the research track I chose in graduate school. Philippine folklorist Damiana Eugenio has this to say about the importance of studying folk literature:

> Rather than destroying folk literature, it is hoped that the greater familiarity of students and city dwellers with our folk literature, which printing is expected to achieve, will preserve its life and will stimulate the collection of more versions of the printed items.[1]

Inspired by the works of a young clergy, Emmanuel Nabayra, who recorded, collected, and aimed to salvage what remains of the ethno-epic of an indigenous community back in the 70s, I decided to retrace his footsteps and find

1 Damiana Eugenio, 'Introduction' to *Philippine Folk Literature: An Anthology* (Quezon City: The University of the Philippines Press, 2007), Second Edition.

out for myself the origin of the oral traditions and their journey towards the printed pages. The research was guided by the goals of folkloric research set by Philippine anthropologist E. Arsenio Manuel:

> (1) To collect and record systematically all oral traditions on the language of each and every ethnic group, (2) To preserve, archive, and classify the materials gathered…, (3) To translate these materials into one or more common languages for study purposes, (4) To publish them for the use or enjoyment of the nation and the world.[2]

That research was a milestone as it became the first of the several studies I conducted with the indigenous people (IP). The goal of the research was to record the chanting of an epic by a native bard. The chanting usually lasts days, and I was there to record the whole thing. The bard shared with me a fascinating and awe-inspiring tale. It told of the adventures of a Bagani, a tribal hero, who saw in his dream the woman he desired to marry. He embarked on an expedition in search of this woman only to find out that she was betrothed to another man. This did not deter him from taking the woman. And so, Paris-like, he abducted the woman and took her away from her groom, which caused a battle to ensue. My fascination was not just on the story itself, but on the whole process of chanting, and most especially on the demeanour of the native bard who was about 70 years old. Nobody else in the village possessed the knowledge and the skill of the bard, and it is a cause of alarm that there are only a few of them left based on the difficulty of finding a resource person while undertaking the researches. Eventually, the research output would later be published in *Tambara*, the academic journal of the Ateneo de Davao University.

2 From Research to Illumination

I continued doing research funded by the University of the Philippines, and determined to cover as much ground as I could in salvaging what still remains of the nearly extinct oral traditions of the ethno-linguistics groups in the country, more particularly in the southern part, in Mindanao, my home island. A pressing reason is the fact that the bards are getting older, and one actually passed on weeks after sharing a significant narrative with me. I became acquainted with the cultures of the Mandaya, the Mansaka, the Blaan, the Bagobo, and the

2 Arsenio E. Manuel, *A Review of Oral Literature Scholarship in the Philippines Universities* (Davao City: Ateneo de Davao University, 1975), Dialogue for Development: Vertical File.

Ata Manobo in the Davao Region. Out of the 112 Philippine ethno-linguistics groups, about 25 are in Mindanao. Indeed there is much work to be done.

To establish authenticity, it is essential to record each oral tradition from the lips of the bard or the storyteller. This gave me the opportunity to closely work with and get to know native resource persons. I was fortunate to get acquainted with 3 Mandaya, 2 Blaan, and 2 Mansaka bards, whom the communities locally call 'magdadawot'.

It was the Mandaya magdadawot who had the most significant impact on me, not only because I was able to visit the area more than once, but also because I was able to spend more time with them. Although I barely understand the language, I was able to communicate with them through a native guide. When I listened to their chanting of the ethno-epics, I could sense their enthusiasm, their passion, and their excitement to be able to share an aspect of their culture to an outsider like me. I felt humbled and honoured at the same time. It was only later that I fully understood the trust and the task that was subtly delegated to me by the bards and their communities.

Academia plays a major role in archiving, preserving, generating, and promoting knowledge. As part of the academe, I was therefore expected and obliged to share the knowledge entrusted to me by the indigenous communities. It was a responsibility I was ready to fulfil. Among my recently published articles based on the research data are: *Yang Mangngagaw: An Epic of the Mandaya* which was published in 1999 in the Tambara journal of the Ateneo de Davao University Journal; *Three Folktales Inspired by the Blaan of Mlang North Cotabato* which was published in 2013 through funding from the National Commission for Culture and the Arts; *Woven Dream, Weaving Stories: Remnants of a Precious Legacy* which was presented and published in the conference proceedings of the International Conference on Arts, Culture, Heritage and the National Development Plan: Vision for 2030 organized by Tshwane University, Pretoria, South Africa in 2014; and *Being Woman: Empowered Matrons and Damsels Not Quite in Distress in Three Blaan Folktales* which was presented and published in the conference proceedings of the International Conference on Arts and Humanities in 2014 in Colombo, Sri Lanka.

My interaction with the indigenous communities allowed me to write and present papers not only based on their folk literature, but on my observation of their daily lives and their experiences. Although my initial goal was to collect and record the nearly extinct oral traditions, I was privileged to see beyond the stories that the bards shared with me. I also saw their own stories. I felt I was able to understand their culture more, and that I must also write about other significant aspects of their culture. *When Gods Cry: Exploring Environmental Crisis Through Literature Inspired by the Indigenous Peoples Experience* was presented in a conference and later published in the Journal of Literature and Art Studies of David Publishing Company in 2013, *Playing the Games, Reliving the Past: Indigenous Games of the Davao Tribes* and FED (*Family, Education, Development*) *for the Women of the Tribes* were both presented in conferences and published in conference proceedings in 2012.

Despite the number of papers I wrote and presented in conferences, I had this nagging feeling that I missed out on a vital aspect of the indigenous people's experiences that I could do more to capture and put into words the fundamental essence of their culture. Although it is true that scientific paper presentations and publications are major venues that promote new knowledge, there is a certain level of objectivity that has to be met. However, this level of objectivity puts a limitation to the scope and mode of expression. And so it dawned on me that I could actually utilize the literary style as a medium for writing about the indigenous people's experiences. It was timely that I came across a book by a Caribbean author, Sharlow Mohammed, titled *When Gods were Slaves*:

> When Gods Were Slaves is a novel dealing with Caribbean history during slavery, that is to say, before Indians first came to the Caribbean in 1838. The action of the novel opens in 1798, near Ile Ife, Nigeria, where a group of Nigerians including Akinjaya, the oba's son, is captured by European slave traders. The scene then shifts briefly to Bristol, one of the British ports that gained most wealth from profits of the slave trade…it provides a comprehensive portrait of plantation life in Trinidad during the first three decades of the 19th century, until Emancipation, when all slaves in British colonies were freed between 1834 and 1838.[3]

I realized I could write using the same style. And so I did.

3 From Illumination to Imagination

As soon as I had time to spare, I went through a flurry of writing stories all of which are an attempt to capture an essential experience of the community.

The first story I wrote is titled *The Last Binukot* (unpublished). It tells about the dilemma of the younger generation to take on the challenge of becoming the repository and the mouthpiece of their oral traditions. Abrera provides a detailed description of this tradition:

3 Frank Birbalsingh, A Review to *When Gods Were Slaves* (USA: Sirens Publications, 2004).

> The female Binukot is isolated by her parents from the rest of the household at 3 or 4 years old. She is not exposed to the sun, not allowed to work, and is even accompanied by her parents when she takes a bath. This results in a fair, frail, fine-complexioned and long-haired woman. As she stays at home most of the time, her parents and grandparents entertain her with various oral lores and traditional dances. This makes the Binukot excellent epic chanters and repositories of their history.[4]

The internal conflict that the young character in the story has to struggle against reflects the present generation's hesitation to be boxed in tradition, and the older generation's apprehension that their traditions will lose continuity in the generations to come. This story was inspired by the lighter moments in-between the chanting of the epics that I shared with the bards, one of whom was a binukot herself until she got married.

After *Binukot*, I wrote *Loom of Dreams* which was published in Krtika Kultura in 2015, a story that got its inspiration from the plight of traditional material cultures, one of which is the *dagmay*, a woven material that itself tells a story. Tradition requires that the dagmay should be revered as a gift of the spirit of nature. The design on this woven fabric is revealed only in a dream, and the weaver tasked to weave it is specifically chosen by the *Tagamaling*, the Spirit of Nature. For a Mandaya, it is therefore sacred. The tension in the story revolves around the struggle to respect tradition and defy the lure of financial gain, a sad reality that is causing a conflict among the indigenous folks.

Invocation which was published in Banwa Online Literary Journal in 2015 is a bildungsroman genre where a young altar boy gains an insight into the seemingly complex relationship between the Catholic Christian religion and the indigenous people's animistic spirituality. As the altar boy of the parish, he was required to accompany the priest to his evangelistic visits to the mountain communities. There he witnessed the tension caused by the IP's suspicions over the real intentions of the Catholic Church, and their apprehension over the eradication of their spirituality. The IPs continue to practice animism despite involvement in various Christian religions who have put up church buildings in many areas of their ancestral domain. In the story, the boy sensed the conflict that ensued between the Catholic priest and the *balyan*, the native priestess, who perceived that the Catholic Church was encroaching in the practice of their faith.

4 Maria Bernadette L. Abrera, 'Seclusion and Veiling of Women', *UP Diliman Journals Online: Philippine Social Sciences Review* 60.1, viewed 7 April 2016.

Touching on a similar theme is the story *Maestro* (unpublished) where a teacher decided to give up his teaching post in a city high school in order to teach indigenous children in a remote mountain school. He was determined to 'civilize' not only the children but also the whole community through the kind of education he was bringing with him. His journey to the mountain brought him in close contact with mythical creatures, the king cobra and the eagle king. In the village, he met the formidable chieftain who made him question his own concept about education and civilization. Who is educating whom? In the brink of death caused by a sullen and suspicious villager, he was brought into the presence of the *balyan*, the village healer and priestess, who saved his life. The story was inspired by my interaction with teachers deployed to the remote villages. They too have to realistically contend with the issue of how far should they go in educating indigenous learners. Their experience became the material for this story.

Waterfall (unpublished) is a novella-length piece inspired by the tragic aftermath of a typhoon that struck a wide area in the southern Philippines in 2013, causing devastation and the loss of numerous lives, many of them from the indigenous communities. The calamity was also the subject of my paper titled *When Gods Cry: Exploring Environmental Crisis Through Literature Inspired by the Indigenous Peoples Experience,* published in the Journal of Literature and Art Studies in 2013. This story highlights the destruction of the natural environment by corporate logging companies and portrays the indigenous people as the protectors of nature. In accordance with their beliefs, the IPs consider natural calamities as caused by the anger of the gods and the spirits. Torrential rains and the typhoons, for instance, are believed to be caused by the incessant crying of the gods. There is then a need to appease the anger of the gods through rituals. Such ritual was performed a couple of days after the destructive typhoon.

Other stories included in the anthology are the following: *Roots,* which deals with how the IPs struggle against droughts; *Lights* which deals with the disenfranchisement of the IPs who were pushed to the hinterlands because of colonization and the effects of urbanization; and *Apo* which is dedicated to the highest mountain peak in the Philippines, home to indigenous communities.

Lights, which is re-titled as *Light Breaks,* has been published in the Banwa Online Literary Journal on December 2016.

4 ...and Finding My Writing Space

I came. I saw. I listened. I understood. I wrote.

This I believe sums up my journey towards writing stories inspired by the indigenous people's experiences. At first, I was a reader and a listener, then a researcher challenged by the intricacies and the fascinating elements of folk literature. As a researcher, I merely recorded data, specifically on indigenous knowledge, and then disseminated this for the consumption and use by researchers and others who find the research output relevant. I assumed this was my niche, my writing space. And then I realized I could do more with the materials and the experience I now carry with me. One by one the stories took form, each a portrayal of a unique and real experience of the IPs.

As I step back and assess the few works of fiction I have written, I ask myself 'Am I now a writer? Have I found my writing space?' I find George Orwell's words 'from a very early age, perhaps the age of five or six, I knew that when I grew up I should be a writer' as both an answer and a question. Unlike him though, I didn't have this realization at a very early age. I would actually consider it quite late to contemplate this question, and quite late to be writing. I should have started years ago. In his essay 'Why I Write', Orwell points out that:

> I do not think one can assess a writer's motives without knowing something of his early development. His subject matter will be determined by the age he lives in ... but before he ever begins to write he will have acquired an emotional attitude from which he will never completely escape. It is his job, no doubt, to discipline his temperament and avoid getting stuck at some immature stage, in some perverse mood...[5]

Perhaps my question should be 'Why am I writing?' and 'What are my motives?' Orwell identified four (4) motives for writing: sheer egoism, aesthetic enthusiasm, historical purpose, and political purpose. I have to strike out sheer egoism, or even aesthetic enthusiasm. The closest perhaps is historical purpose, as it entails the 'Desire to see things as they are, to find out true facts and store them up for the use of posterity', and a bit of political purpose that expresses 'Desire to push the world in a certain direction, to alter other peoples' idea of the kind of society that they should strive after' I do have these 'desires' to write for posterity and to 'push' my readers into seeing the real plight of a certain sector in society. Reading my stories now, I actually feel relieved that I got to write them. I am now beginning to understand the trust that the magdadawots bestowed upon me . I realize now that they have indirectly tasked me to write for them, not just about them, and to be some sort of a mouthpiece for them.

5 George Orwell, 'Why I Write'. Viewed 7 April 2016, http://orwell.ru/library/essays/wiw/english/e_wiw

I felt like the hesitant binukot in *The Last Binukot*, because at first I didn't quite comprehend what was expected of me. I was conflicted. After all, it is a great responsibility to write for and about the indigenous people's experiences. I felt like the young altar boy in *Invocation*, a mere observer, naïve at first but later drawn into an insightful experience and emerging wiser. I felt like the young teacher in *Maestro*, wanting at first to 'educate' the IPs, but ended up being educated by them.

Empirical research requires that new knowledge is written down and disseminated. Literary writing requires research, and perhaps immersion, so that the writer gains a clearer foundation for his or her narratives.

It has been a compelling journey for me. I started out just being fascinated by the stories of old, and when the curiosity set in, I set out in search of the origins of these stories and found more than I expected and was ready to find. I met significant people along the way who showed me what I could do. I was drawn into the life and the experiences of indigenous communities, their passions, their conflicts, their struggle against natural and external social forces, their need to keep their traditions alive, their determination to keep their cultural identity, and their need to be heard. I would like to think that I have evolved. I was a reader, a researcher, and now, I would like to believe, a writer.

I have found my writing space.

Bibliography

Abrera, Maria Bernadette L. 'Seclusion and Veiling of Women'. *UP Diliman Journals Online: Philippine Social Sciences Review* 60.1. Viewed 7 April 2016 http://journals.upd.edu.ph.

Bresnahan, Roger J. *Literature and Society: Cross-Cultural Perspectives*, 2nd Edition, Eleventh American Studies Seminar, Los Baños, Philippines October 1976.

Castro, Jovito V. Contributor. *Antolohiya ng mga Panitikang Asean: Mga Epiko ng Pilipinas*: Quezon City: APO Production Unit Inc., 1984.

Coronel, Maria Delia. *Stories and Legends from Filipino Folklore*: Manila: University of Santo Tomas Press, 1966.

Eugenio, Damiana. *Introduction: Philippine Folk Literature: An Anthology*. 2nd Edition. Quezon City: The University of the Philippines Press, 2007.

Jocano, F. Landa. *Outline of Philippine Mythology*. Manila: Centro Escolar University Research and Development Center 1969.

Manuel, Arsenio E. *A Review of Oral Literature Scholarship in the Philippines Universities*. Dialogue for Development: Vertical File, Davao City: Ateneo de Davao University, 1975.

Mohammed, Sharlow. *When Gods Were Slaves.* USA: Sirens Publications, 2004.

Orwell, George. *Why I Write.* Viewed 7 April 2016. http://orwell.ru/library/essays/wiw/english/e_wiw.

Quintero, Genevieve J. 'Invocation'. *Banwa Series A* 12. Ateneo de Manila University, 2015. Viewed 8 January 2016. http://ojs.upmin.edu.ph/index.php/banwa-a/issue/view/23.

Quintero, Genevieve J. 'Loom of Dreams'. *Kritika Kultura* 25. Ateneo de Manila University, 2015. Viewed 8 April 2016. http://journals.ateneo.edu/ojs/kk/article/view/2192/2187.

Quintero, Genevieve J. 'When Gods Cry: Exploring Environmental Crisis Through Literature Inspired by the Indigenous Peoples Experience'. Journal of Literature and Art Studies. Issue 10. ISSN 2159–5836. USA: David Publishing Company, 2013.

CHAPTER 3

Attachment or Antithesis? Middle School Children and Writing as Relational Consciousness

Phil Fitzsimmons and Edie Lanphar

Abstract

Using a bricolage of design and methodologies, the research findings unpacked in this paper emerged out of a qualitative project that sought to illuminate a key aspect of the 'four resource model' of writing. This critical point of speculative focus was the notion of the habitus-writing connection, and how, if at all, it arose within the writing of twenty-six grade six students in one Catholic school in northern California. The 'four resource model' takes up Bordieu's[1] concept of habitus as being internalized pre-dispositions, which are the 'product of a shared history in a child's family'[2] In regard to writing, it has been ventured that young children are subject to both the home life literary influences, as well as the literary aspects within their socio-cultural spheres. However, these influences are not set in cognitive-concrete, as they also appear to place their own perspectives into the texts they create: innovating and exploring as they write. While initial interrogation of the data revealed evidence of a habitus related to genre, ensuing 'focused coding' revealed evidence of a secondary layer of reflexivity underpinning the surface features of their texts. This layer of meaning making had characteristics of a melding of attachment and the spiritual notion of 'relational consciousness.' It would appear that, for these children at least, a key component of personal writing is a connectivity grounded in relationships.

Keywords

Attachment – spirituality – writing – middle school

•••

1 Pierre Bordieu, *An Invitation to Reflexive Sociology* (Chicago: University of Chicago, 1992).
2 Pauline Harris, Jan Turbill, Phil Fitzsimmons and Barbra McKenzie, *Writing in the Primary Years* (Sydney: Cengage, 2001), 46.

 | DOI:10.1163/9789004394315_004

1 Locating This Study in Process and Place

This project reports on findings associated with understanding how one cohort of year students reflectively responded through writing to a reading-writing connection[3] project that was a key component of their daily literacy activity. As will be briefly discussed later in this chapter, this project was grounded within a long held psycho-social-linguistic or natural learning perspective, which takes the axiomatic stance that all aspects of literacy development can be viewed as language development. That is, reading, writing, speaking and listening are learnt in the same way and through similar stages. Hence, in school literacy learning, these aspects should be taught as a unified whole as they each support one another. For many researchers in this field these ideals emerged out of the work of Brian Camborne's[4] who believes that if children are provided with powerful and engaging holistic literacy demonstrations, and allowed to take responsibility for their own learning, then authentic engagement will occur. As we see the term and ensuing practice, holism is about developing children's inner and external dialogue so they 'they read like writers and write like readers.'[5] While this has been accepted proposition in the early years of a child's development, the question remains as to why this process should stop or be watered down in the latter years of school? Given the issues of falling literacy rates in general in secondary school sectors,[6] and student's apparent inability to write in particular we have begun to question the current curricula regimes of many educational authorities that have focused on constraint and conformity. We have not been alone in this questioning, as researchers such as Carter[7] have questioned why many politicians feel the teaching of writing should 'be controlled by a more regimented curriculum for English with decontextualized grammar drills and proper, standard English.'

While the conditions under which natural learning have been clearly identified, the overarching thought processes in which they are embedded have not been clarified. As well, little research appears to exist in regards to how older

3 Jerome Harste, Kathy Short and Carolyn Burke, *Creating Classrooms for Authors: The Reading-Writing Connection* (Portsmouth: Heinneman, 1988).

4 Brian Cambourne, *The Whole Story: Natural Learning and the Acquisition of Literacy in the Classroom* (New York: Scholastic, 1993).

5 Frank Smith, *Essays Into Literacy: Selected Papers and Some Afterthoughts* (Porstmouth: Heinnerman, 1983), 126.

6 Daniel Perrin and Sue Graham, *Writing Next: Effective Strategies to Improve Writing of Adolescents in Middle and High Schools – A Report to the Carnegie Corporation of New York* (Washington, DC.: Alliance for Excellence in Education, 2007).

7 Ron Carter, *Language and Creativity: The Art of Common Talk* (London: Routledge, 2005), 12.

students respond to a classroom situation in which holistic learning has been implemented. While acknowledging the small scale of the project in focus in this chapter, the aim of this project was to step into minds of children to gain the most elementary awareness of their perspectives.

As with many qualitative research projects, the underpinning paradigmatic praxis of 'emergent design' almost ensures that the 'smart bomb' interaction with data and respondents produces new directions of understanding. The project discussed in this paper is no different and eventually morphed into three conjoint research foci with this particular paper focussing on children's understanding and approach to writing. Data were collected through a bricolage of semi-structured interviews, in-situ observations, and ad hoc discussions with the student respondents and the reflective journals generated by each class member. Indeed, as the project emerged, the latter became the primary focus of all the data collection and analysis.

With an initial focus of seeking to explore the nature of any habitus or memory connections that emerged in the writing of a class of twenty-eight year six students in one Catholic school in California, the 'three level coding'[8] process of analysis and constant comparison across all data sets revealed blips of data that we believed were related to the notions of spirituality and attachment. Before moving further into this chapter we need to make it clear that our axiomatic position that habits and memory connections, can be viewed synonymously in that,

> The mechanism of memory system tells us that we do not have raw sensation. Sensations that we experience are the aftermath of a filtering process through feelings, values, attitudes, beliefs, value systems, memories and self narrative.[9]

In digging deeper into this data set, and ensuring that the 'eye of the beholder' syndrome was not generating our initial hunches, or was a product of tacit awareness, we came to realize through the triangulation of data that these blips were in fact not mediated only through our *insider/emic perspective* but could be shown to be propositional constructs. As will be discussed in an ensuing section, what initially appeared as blips were in fact emergent elements that were connected to a deeper aspect of children's reflective capabilities: something akin to an iceberg.

8 Kathy Charmaz, *Constructing Grounded Theory*, 2nd ed. (London: SAGE, 2014), 115.

9 Sangeetha Menon. *Brain, Self and Consciousness: Explaining the Conspiracy of Experience* (London: Springer, 2014), 11.

This iceberg of reflexivity arose in a project-based learning[10] exercise that had its overall and daily genesis in a read aloud or shared book experience.[11] Using Lois Lowry's *Gossamer,* this text was read to the class each day, which then evolved into a whole class reflective discussion. This young adult fantasy-fairy tale text was chosen as it represented not only quality literature, but also appeared to have the potential to provide a series of touchstone points that had the potential to not only stimulate discussion but also provide an entrée into the deeper memories and personal connections these emerging adolescents may hold.

> While it entertains the child, the fairy tale enlightens him about himself, and fosters his personality development. It offers meaning on so many different levels, and enriches the child's existence in so many ways, that no one book can do justice to the multitude and diversity of the contributions such tales make to the child's life.[12]

From the reflective discussion, key points of wording, sentence construction and paragraph connectivity were used as models of writing that invited or engaged these students to 'develop a relationship with the text.'[13] The students were invited to consider how to incorporate similar forms of language use into their own writing. These were often shared and discussed further in the final sharing episode of each lesson.

2 Locating This Study in the Research and School Contexts

The overall context of the data that forms the basis of this paper has arisen out of the American and Australian educational contexts. In the pendulum swings of curricula and apparent political forces in these countries in the past two decades, it seems to us that much of what was learnt from previous ethnographic research with children and research into teaching practice in the previous

---

10 Margot Filipenko, Jo-Anne Naslund and Linda Siegel, 'Discovering, Unearthing and Creating Meanings: Problem Based Learning in Teacher Education', *Problem Based Learning in Teacher Education,* eds. Margot Filipenko and Jo-Anne Naslund (London: Springer, 2016), 1–10.

11 Allan Trent and Jeasik Cho, 'Evaluating Qualitative Research', *Oxford Handbook on Qualitative Research,* ed. Patricia Leavy (New York: The Oxford University Press, 2014), 677–696.

12 Bruno Bertelheim, *The Uses of Enchantment* (New York: Vintage), 173.

13 Louise Rosenblatt. 'Towards a Transactional Theory of Reading', *Journal of Literacy Research* 1,1, 1969: 35.

decades has been lost. While we are acutely aware that educational praxis is never linear or hierarchical, in this instance the key component of what actually writing represents has been subsumed by a view that is reminiscent of much earlier twentieth century viewpoints. In other words, the paradigmatic regression has been much more severe in this period of time. Perhaps more importantly, while the political masters in both countries have demanded research 'evidence based,' what is really meant is 'show us the numbers,' and if we can't measure it, it does not matter.

While there are obviously gifted teachers of writing in the primary or elementary years of school in Australia and the United States, it has been our experience that these kinds of teachers are becoming increasingly rare. As stated, the teaching of writing has gone through radical changes in the last decades, because writing in this era is viewed as grammatical structures, grammatical linkages and grammatical purposes. While the United States has not taken on board the systemic functional approach of the Australian scene, it seems that writing has become increasingly focussed on essay writing in that context. As Jesson, Parr and McNaughton frankly admits, there appears to a wide reaching disconnection between the language use and literacy practices of many students, and many school practices of focussing on a text type that does not resonate with students understanding of semantic application or grammatical construction. 'There is evidence supporting this disconnection hypothesis; that standard practices of schooling preclude some students' knowledge and ways of participating.'[14] Both overarching systems have become focussed on writing as genre or text type as the belief is the overarching concern has been that children are not being prepared for essay writing or factual writing in high school. While we also concur with this assertion to some degree, it's our opinion that the curriculum pendulum swing has thrown the baby out with the bathwater. In other words, in the shift towards a functional grammatical or structural approach how writing surfaces naturally before school has been ignored. The literacy wars of the 1990's also subsumed the most basic understandings of how writing developed, and began to shatter many of the linkages regarding the reading-writing connection in which it had become enmeshed.

In regard to the latter point, another critical aspect is the actual volume of competing research paradigms related to writing. Research into writing development has a long history, however as Bouffler stated three decades ago, 'it is

14 Rebecca Jesson, Judy Parr and Stuart McNaughton, 'The Unfulfilled Pedagogical Promise of the Dialogic in Writing: Intertextual Writing Instruction for Diverse Settings', *International Handbook on the Researrh on Children's Literacy, Learning and Culture,* eds. Kathy Hall, Teresa Cremin, Barbara Comber and Luis Moll (Oxford: Wiley-Blackwell, 2013), 216.

difficult to make sense of the plethora of material available. It is even more difficult to relate it to a consistent view of language.'[15] This project is specifically located within what we believe is a 'consistent view of language' development, which is grounded in a psycho-social-linguistic perspective. Drawing on ethnographic research this perspective takes the axiomatic stance that children bring a great deal of experience and understanding to all aspects of their literacy development. Not only are children able to gain control over the secretarial aspects of writing for example, but also engagement in writing has the potential to be an important component of a child's overall development and one of the ways in which children are able to externalize their inner thoughts, emotions and begin the journey in developing a focused worldview. This has become the antithesis of many school settings in the Western World.

However, while writing alone can provide a more focused awareness, the use of a reading-writing connection provides a broader and more connected platform of awareness. This often begins in the home well before the start of school in which their parents or carers read to their children, and gradually allow them to begin to engage in reflective questioning and writing. Hence, both reading and writing becomes an integral component of a relationship and not an isolated event. As well, if they are provided with any tools of writing, they also begin to engage in a multi- modal process of writing, thinking and further engaged talk with their significant other. Kress believes that this interactive and multi-modal process reflects a deep thinking process that is both unconscious and conscious switching of emotions. This process he believes 'is essential for human beings to understand the world.'[16] It seems that this is an easy process for young children who typically do not have the pressure to conform to standard spelling or grammar as their main concern is message formulation and the co-construction of identity.[17] Thus, all their energies are drawn from their own knowledge, environmental sources or simple invention. Researchers in this field believes the entire writing process for children is one of refinement: the more they write and take risks, the more they are refining their understanding of print and their broader world view. However, as Mavers points out, children engaged in writing are actually 'semiotic workers – with a working identity.'[18] It is only in school that this process of writing often gives

15 Christine Bouffler, *Spelling: A New Approach to An Old Problem* (New York: Scholastic, 1992), 86.

16 Gunther Kress, *Before Writing: Rethinking the Paths to Literacy* (New York: Routledge, 1997), xv.

17 Richard Mayer, *Multimedia Learning: Second Edition* (New York: Cambridge University Press, 2009).

18 Ibid., 9.

way to 'inadequacy and deficit'[19] where as the basis of writing before school is one of active agency.

Much of the research data regarding writing in the ethnographic literacy research space has arisen out of observation, analysis of writing and interviews regarding children's perspectives. While valuable in gaining understanding regarding how literacy is developed, Kress has argued that while holistic in nature, 'we need a much more generous understanding of cognition than the sparse understanding we have now—and one that does not sever pleasure from thought, and thought from emotion, feeling, affect.'[20] Beyond references to writing development being linked to identity formation, our review of the current literature suggests that there is still a great deal of research that needs to be done in this area.

3 Locating This Study in Data and School Context

Using Charmaz's[21] three level system of coding qualitative data several critical aspects emerged from within the data. While editorial limitations preclude the detailing of all the emerging themes, the nested core aspects of the data analysis included:

3.1 *Connecting With Emotional Linkages*

While the student respondents in this project discussed the notion of reflection, it was interesting to see in their reflective responses a general overall transition in all their writing a shift from retell to thinking through to feeling. It should be noted that while the children were unable to describe how this came about, it was certainly noted that this was the pattern of interaction and questioning that came into play in the reflective discussions that arose from the text. It could be that the nature of the focused reflection became a sublimated facet of the student's reflection, which was then mirrored in their writing.

In regard to emotion, the students reported that this has been for the most part discouraged in their school experience and that the emphasis on factual reporting through an essay model or text type had 'only gotten one part of me in print.'[22] While obviously revealing the student's desire to be able to include

19 Ibid.

20 Gunther Kress, *Literacy in the New Media Age* (New York: Routledge, 2003), xv.

21 Kathy Charmaz, *Constructing Grounded Theory* (London: SAGE, 2014), 19.

22 Student A, Interview with first author.

and unpack their emotional reactions, if not their emotional self, this shift also divulges the possible use of writing as the modality through which expression of self is equally if not more important than the communication of supposed objective understanding. These children reported writing in many instances was simply a set of skills to be practiced rather than an opportunity to fully explore reactions and responses. Interestingly, many of these students engaged in writing at home, and other forms of expression such as drawing and computer graphics. They were also interested in other diverse genres such as film and photography. However, it was only in this class that they were able to cross and intertwine these semiotic forms and engage with their teacher and peers in what appears to be an authentic interactive and reflective undertaking. Writing allowed the majority of these students to reflect and touch darker aspects of the childhood, or as Student E stated, visit a 'day of darkness.' Rather than being a negative, writing provided the mechanism through which these students could 'put it all into one little square. I wrote it, then I could look at it.'[23]

As indicated by the previous data, for these children, writing as reflection in this instance seems to have been the means through these aspects of multimodality and connectivity also crossed the internal political or power based borderlands of the mind. As student F wrote: 'Have you ever not liked your parents for being grounded?'[24] He then went on, as many others did, of using writing as a release of honest tension or perceived tension between the student's notions of them and us, interlocking the students' ideals of adults, teachers and parents with their sense of self. This is exemplified in an impromptu discussion between the first author and Student D as the latter was writing:

> Student D has written: Fear, joy, anxiousness, and happiness, all of those things. The most minor things in life are sometimes the ones that hurt the most when you lose them.
> P: What's going on as you are writing your reflection?
> Student D: I am really having to think about who I am? It's not easy, the whole writing is not easy but I am seeing stuff, new stuff.[25]

3.2 *Connecting with Familial Linkages*

Integrally linked to the previous theme, the data supplied by these respondents unearthed a series of references that clearly identified a deep emotional

23 Student F, Interview with first author.

24 Ibid.

25 Student D, Interview with first author.

attachment to family members, even though some of these children appeared to be some form of dire familial circumstances. The majority of children in this study reported that this was the first time they had been able to express their feelings towards others and that listening to the text and other classroom members not only allowed them to see things in the book they had not imagined, but also 'made me think about me and what things really meant to me.'[26] While perhaps not entirely surprising, these elements are not entirely consistent with the view of middle school children as viewed by many in the general populace. In our view, it clearly suggests that children in this age bracket still desire and need a close emotional connection with their parents and extended family. Indeed, the memories that surfaced in their reflective writing clearly indicate that writing in this instance was an enabler in allowing these students 'to realize what things we have in our life that shape who we are.'[27] While the descriptions these students supplied were tinged with nostalgia, it was realized as a positive emotion and linked with mostly positive memories. More importantly, all of the descriptions and recollections were connected within a web of closeness to others. More importantly these reflections were also couched in a high modality of language such as the reflection supplied by Student B:

> It reminds of my dad and how my family is always there when I need them or to help me correct a mistake that I have made. I have had such good memories playing sports.[28]

As well, there was a general pattern in the writing that often shifted from firstly negative instances to positive aspects. It seems that this reflective-writing instance allow familiar connections to surface in a simple but in apparent deeply embedded manner.

3.3 *Connecting With Outside of Self Linkages*

Embedded throughout these reflective pieces, the many students referred to religious or relational aspects beyond themselves. As Student E wrote: 'It taught me a lot and made me think about life and the people around me. It also made me think about what I can do that can influence the people around me and make their life better.'[29] While there is an apparent relationship to the text

26 Student L, Interview with first author.
27 Student N, Interview with first author.
28 Student L, Interview with first author.
29 Student E, Interview with first author

they had been reading together, this was not the main influence as suggested by Student F while engaged in the writing process.

> It's easy to talk about a book and make those kind of,... instant connections. But when you have to write it's like water going down a sink. Everything gets pushed in together in a swirl and you have to make it all go slower down the sink. It makes you think outside of my head. Way outside.[30]

4 Locating the Study in the Threads of Data and Responsive Research

This small scale project suggests that writing could be both a mediating and an organizing force in regard to middle school students understanding of self and their personal potential. In this instance writing became an emotional trigger, or a form of an emotional 'can opener.' In a final responsive comparison of the findings with the broader research literature, it became clear to us that reflective writing is not simply a mechanistic means to an end, but allows the development of a conjoined and holistic understanding related to spirituality and attachment. In the broader research sphere, spirituality and attachment are both concerned with connection and relationships. As Hay and Nye have consistently asserted children's spirituality involves an awareness of one's relationship with self, with others and the entirety of the world around them. They have argued that such an 'awareness of this holistic relationship with the rest of reality is central to the nature of spirituality.'[31] Writing, within a reading-writing connection could be a means through which both teachers and children gain insight into this spiritual-attachment connection.

Bibliography

Bertelheim, Bruno. *The Uses of Enchantment.* New York: Vintage, 1992.

Bordieu, Pierre. *An Invitation to Reflexive Sociology.* Chicago: University of Chicago, 1992.

Cambourne, Brian. *The Whole Story: Natural Learning and the Acquisition of Literacy in the Classroom.* New York: Scholastic, 1993.

30 Student F, Interview with first author.

31 David Hay and Rebecca Nye, *The Spirit of the Child* (London: Harper Collins, 1998), 142.

Bouffler, Christine. *Spelling: A New Approach to an Old Problem.* New York: Scholastic, 1992.

Carter, Ron. *Language and Creativity: The Art of Common Talk.* London: Routledge, 2005.

Charmaz, Kathy. *Constructing Grounded Theory*, 2nd ed. London: SAGE, 2014.

Harris, Pauline, Jan Turbill, Phil Fitzsimmons and Barbra McKenzie. *Writing in the Primary Years.* Sydney: Cengage, 2001.

Harste, Jerome, Kathy Short, and Carolyn Burke. *Creating Classrooms for Authors: The Reading-Writing Connection.* Portsmouth: Heinneman, 1988.

Hay, David and Rebecca Nye. *The Spirit of the Child.* London: Harper Collins, 1998.

Jesson, Rebecca, Judy Parr and Stuart McNaughton, 'The Unfulfilled Pedagogical Promise of the Dialogic in Writing: Intertextual Writing Instruction for Diverse Settings,' *International Handbook on the Research on Children's Literacy, Learning and Culture*, 255-228, eds. Kathy Hall, Teresa Cremin, Barbara Comber and Luis Moll Oxford: Wiley-Blackwell, 2013.

Kress, Gunther. *Literacy in the New Media Age.* New York: Routledge, 2003.

Mayer, Richard. *Multimedia Learning: Second Edition.* New York: Cambridge University Press, 2009.

Menon, Sangeetha. *Brain, Self and Consciousness: Explaining the Conspiracy of Experience.* London: Springer, 2014.

Perrin, Daniel and Sue Graham. *Writing Next: Effective Strategies to Improve Writing of Adolescents in Middle and High Schools – A Report to the Carnegie Corporation of New York*: Washington, DC.: Alliance for Excellence in Education, 2007.

Rosenblatt, Louise. 'Towards a Transactional Theory of Reading,' *Journal of Literacy Research* 1,1, 1969: 31-49.

Smith, Frank. *Essays Into Literacy: Selected Papers and Some Afterthoughts.* Porstmouth: Heinnerman, 1983.

Trent, Allan and Jeasik Cho. 'Evaluating Qualitative Research,' *Oxford Handbook on Qualitative Research. The Oxford Handbook of Qualitative Research*, edited by Patricia Leavy, 639-657. New York: The Oxford University Press, 2014.

CHAPTER 4

Breathing In; Breathing Out: Writing as a Spiritual Space

Gail Hammill

Abstract

Conventional houses of worship, whether mosques, churches or temples, rarely call forth spiritual sensibilities in me. Such spaces, on the contrary, usually cause me to feel, especially as a woman, overshadowed, barred, swept aside. In essence, stifled. I *have* felt spiritual stirrings in a number of other places such as a museum, a ballet class, or the halls of my children's school. But such stirrings steal upon me at the oddest times and disappear too quickly into the ether. Unless, I write. Dictionaries usually define the spiritual first as something airy or incorporeal, and only secondarily as related to religion. We live in a time when many people in the Western world are turning away from organized religion without necessarily turning away from spirituality, though just what the spiritual *is*, can be difficult to define. Recently, defining the spiritual has become easier for me because I've begun to see writing as a spiritual space. My chapter explores daily writing as a spiritual commitment, drawing on the origins of the word *spirit* which derive from the Latin *spiritus*, meaning 'breathing, breath, or air'. I consider ways that writing can be likened to breathing, in terms of withdrawing from the world and releasing or giving back to it. In so doing I explore reasons for the formidable trials that come along with it. For example, given that I now conceive of writing as a sacred space, why do I still frequently resist going there? Why am I so better at inspiring and having faith in my students than I am at inspiring and having faith in myself? Finally, how is enduring and overcoming such trials dependent upon literary communion with others?

Keywords

Writing – spirituality – spaces – gender – community – teaching

•••

 | DOI:10.1163/9789004394315_005

The modern church where my great aunt Peg's funeral was held was a stark contrast to the ornate Gothic reproduction that stood at the top of my Grandmother's street, a grey-stone edifice where, sitting in the creaky wooden pews in the dim light of the votive candles, you could get lost contemplating the stories in statuary and stained glass. At the funeral, the altar wasn't raised on a marble platform like at my Grandma's church; it was a minimalist affair, designated merely by a scarlet carpet where it stood at the centre, flanked by vases stoked with lilies. The priest, a bald man probably in his sixties, clad in his Catholic lace trim, was assisted by two altar boys about age thirteen, one of whom had reddish-blonde hair and, judging by his facial features, appeared to have Down syndrome. The boys wore the plain white robes over longer black ones beneath which their high-top sneakers anachronistically peeked out, a reminder that the Catholic Church was a champ at the game of survival. Also assisting the mass was a diminutive, elderly nun who, except for the black veil secured to her short grey hair, looked more modern than her male counterparts. Her powder-blue blouse was neatly tucked into a dark blue straight skirt and she wore comfortable, no-nonsense beige loafers.

I had not known my Aunt Peg very well, so these prosaic surroundings were enough to set my mind wandering to consider such things as what bill I should pay next and how many calories I might get away with at the wake. My thoughts eventually circled around again to the robed figures in front of me where the nimble movements of the nun curiously begged attention. Assisting the earthly male trinity as they prepared for communion, she moved with the grace of a sylph. Her balletic bravura seemed uncalled for until I realized that her intricate footwork resulted from the immense circumspection required to avoid accidentally stepping into the altar space. This sacred area was barred to her delicate feet because she was a woman. Two prepubescent boys could tromp sheepishly back and forth across this space in dog-eared high-tops but she had to take the utmost care, in her beige loafers, not to.

It was disturbing to see the nun so assiduously assisting in excluding herself from an institution to which she had devoted her life. Probably she saw it differently. Like my aunt, she was of another time. Unlike them, I was given religion only half-heartedly by my parents who had left the church themselves but sent me to weekly catechism classes until I reached eighth grade. Not surprisingly, without any reinforcement at home the education never took. I've rarely felt spiritual in a church, the one exception being when people were getting married. Most other times, being in a church made me feel, unconsciously first as a girl, then consciously as a woman, like my existence was an aside. Certainly, it never awakened anything in me.

Most dictionaries define the spiritual first as something airy or incorporeal and only secondarily as related to religion. The noun *spirit* has a greater variety of meanings indicating the inner or immaterial quality of a person—the character, essence or soul—along with concepts of that essence surviving after death sometimes in the form of an apparition.[1] Today, many people, especially in the Western world, are turning away from organized religion without necessarily turning away from spirituality, though just what the spiritual is, can be difficult to define.

Fortunately, though I've not found spirituality in religion I have sensed it elsewhere. One place I find spirituality is in words, spoken or written, and especially their etymologies. I look to the *Oxford English Dictionary* as the Bible of the English language where many revelations can be found. I've come to have an almost superstitious faith in the wisdom a word's evolution imparts. This holy book tells us that the word *spiritual* derives from Anglo and Old French versions of the word as well as Medieval ecclesiastical use of the Latin *spiritus*, all of which mean 'breathing, breath, air'.[2] The association of *spiritus* with air which some Latin dictionaries extend to *wind* or *breeze* is puzzling at first, but I think it exists because of the sensation of arousal that accompanies the spirit's stoking. It can sometimes feel like the gentle caress of a breeze while at other times it can feel stronger, like a wave of wind or energy was passing through you.

I don't have this feeling very often. In fact, it even seems to depend on your not expecting it, like it has to take you unaware. Hoping to meet it seems almost always to cause it to flee into hiding. I've been taken unaware in any number of places. Once it happened at the Metropolitan Art museum because the dreamy black eyes of a young woman in a John Singer Sargent painting took hold of mine; another time I felt it standing in the sand by the Pacific Ocean when at dusk my sister and I were gathering the kids to go home and suddenly everyone involuntarily stood to attention because someone spotted dolphins arching over the horizon; I've even felt it a couple of times at my kids' small international school in Dubai where the halls echo with a blend of children's and teachers' voices and the walls are plastered with art projects and photographs of children all wearing the same blue uniforms but having every colour of skin. But the problem is these spiritual moments are too infrequent for my liking and they disappear too quickly into the ether before I can catch them.

Unless, I write.

A few years ago, I began writing personal essays as an attempt to retrieve this ephemera. Sometimes only managing twenty minutes a day, I've come to

1 *Oxford Dictionaries*. "Spirit". Accessed January 15, 2017. https://en.oxforddictionaries.com/definition/spirit.

2 *Oxford English Dictionary*, vol. 10 (London: Oxford University Press), 617.

think of this time spent writing memoir as a spiritual space. The writing of memories offers me a space where I feel fully present in a way that I never could in any church. Memories, I've found, can be a powerful conduit of the spirit. Sigmund Freud wrote eloquently in *Totem and Taboo* about the power of memory to conjure spirits:

> When we, no less than primitive man, project something into external reality, what is happening must surely be this; we are recognizing the existence of two states—one in which something is directly given to the senses and to consciousness (that is *present* to them), and alongside it another, in which the same thing is *latent* but capable of re-appearing. In short, we are recognizing the co-existence of perception and memory, or, putting it more generally, the existence of *unconscious* mental processes alongside the *conscious* ones. It might be said that in the last analysis the 'spirit' of persons or things comes down to their capacity to be remembered and imagined after perception of them has ceased (italics Freud's).[3]

Freud emphasizes here that memories, immaterial as they are, originate in sensory, embodied experience that we desire to have repeated; that the only way to approximate this repetition is to imagine it, which helps to explain why memories tend both to titillate and frustrate desire at the same time. Sometimes when I find myself lost in memories I feel desperate that I can't make them reappear, that I must sort through shadows of perceptions rather than real ones. Sometimes I want to go back to relive and savour rare victories or blissful moments; other times I want to revisit the scenes of my youth but remain invisible like Scrooge in his Christmas dream.[4] I want to take it a step further though and be able to coach myself: *What are you doing? You're an idiot. Why don't you try this? Who cares? Forget about it.* Other times I'm desperate to go back to be certain about what really happened. *Did he say that or did I wish it into memory? Did that happen before or after?* Writing is an attempt to project those moments and nail them down, and even though I never do, I am left with something else when I finish an essay that though never fully satisfactory, pays tribute as a memento. My experiences have not been entirely lost because I have made something from them. Chiselling at this memory-art can be compared to praying because in setting down parts of my life I am in some way giving thanks for them. Van Gogh conveyed something similar in his letters:

---

3 Sigmund Freud, *Totem and Taboo* 1912–1913, trans. James Stratchey (New York: Norton, 1989), 117.

4 Charles Dickens, *A Christmas Carol*.

> I have a certain obligation and duty ... as it were—because I've walked the earth for 30 years—to leave a certain souvenir in the form of drawings or paintings in gratitude (underlines van Gogh's).[5]

And yet, unfortunately, recognizing writing as a spiritual space hasn't made this haven any easier for me to visit. Why do I continue to remain so skilled at avoiding going to this space on a daily basis? Granted, like most people, I'm busy with work to make a living and tired because of it. But why do I often convince myself that I can't even spare just twenty minutes? Why does each day missed make me *less*, not more, inclined to go back? And why am I so much better at inspiring my students to write than I am at inspiring myself? Once, I actually had a student standing in front of my desk who interrupted me to say, almost as if he had an urgent need to go to the bathroom, 'You've just given me some ideas and I'm feeling like I need to start writing them right now'. By all means, I thought, but how did I even get that to happen? And why can I not spark some urgency in myself instead of having to plod hang-dog to the computer every time?

Entering the space of writing, for me, is about as inviting as getting into a cold swimming pool on a rainy day. I simply don't want to get in. Diving in is an option only when I have a pressing deadline. Otherwise, I have to coax myself away from the world by writing first about something I've read. This allows me to eventually slip in with the least pain: feet, then ankles, and so on like a damn coward. Even once I'm in, it takes me a while before I'm really swimming. I have to keep reassuring myself that the next paragraph will definitely be an easier one even though it almost never is by the time I get there.

Eventually though, with writing as with many things, especially when you have the time you really need—an hour or two, you move beyond, not the difficulty, but the caring so much about it. Like a person given morphine, the pain is still there but you find yourself hovering somewhere outside it. Then you've become that other definition of spiritual in the sense of leaving the body, of becoming an apparition. Shedding the body, you travel like the people in science fiction movies moving back and forth in time and to alternate universes. You reinhabit your dreams and former selves; and sometimes other people's dreams and former selves. A few times while writing I ventured so far from home that I let out a scream when my husband spoke to me so startled was I to return to our living-room.

---

5 Vincent van Gogh, '371 To Theo van Gogh. The Hague, on or about Tuesday', 7 August 1883, *Vincent van Gogh the Letters, van Gogh Museum*, Viewed 6 June, 2016, http://vangoghletters.org/vg/letters/let371/letter.html.

Like with swimming, once I've been at it awhile I wonder why getting in was so difficult. What was all that resistance about? Especially since if some travel agency were advertising trips to the past or to alternate universes I would be first in line to get my ticket. But, of course, that's fantasy. In the real world, perhaps the leaving of one's body and day-to-day life, whether through writing or some other activity, requires the withstanding of psychological g-forces. Maybe that's why we don't always feel up to it.

Part of this sluggishness no doubt has to do with the great rigor writing calls for that, like physical exercise, can be daunting at the outset because of the great effort and stamina required. Still, there's another aspect of the origins of the word *spiritual* to consider—the association with breath and breathing. At the very least this connection helps to explain why I've felt my being stifled in a church and at peace in my yoga class. Considered in more depth, however, it brings to mind a beautiful passage from an essay by Tamim Ansary in which he reflects on the role of prayer in the Islamic way of life in the Afghan village where he grew up during the 1950s:

> We didn't spend much time pondering Islam. We didn't have to. Islam permeated the life of the compound like the custard that binds a casserole together, hardly separable from everyday life. Five times a day, some of us did our ablutions and moved into the prayer ritual, one by one, at our own pace. Prayer divided a day into five parts and gave a sort of rhythm to the household, like breathing in, holding for quiet, and then breathing out, releasing back into noise and activity.[6]

One of the many wonderful things about this passage is how, in comparing prayer to respiration, Ansary evokes not only the repetition, and regularity of prayer, but also the necessity and naturalness of it. And yet one cannot forget when reading the passage that this naturalness is achieved only with discipline, the requirement of five daily prayers being one of the main pillars of Islam. Writing is like this. It requires the same disciplined commitment, the same temporary retreat from life as usual, and from other people. It involves a stopping though the current rushes on, and then a releasing of oneself back into the current.

The act of writing, a kind of prayer, requires the same withdraw from the community and the same flowing back to it. Like breathing, the two are not

6 Tamim Ansary, 'Villages and Compounds', *The New World Reader: Thinking and Writing about the Global Community*, ed. Gilbert H. Muller (New York: Houghton Mifflin Company, 2005), 439.

opposites, but part of the same process. The Islamic community rendered in Ansary's portrait respectfully supports each individual's temporary withdraw from itself. A writer's community has the same vital importance. A few years ago, before I'd fully realized this, I attempted to convince myself to write daily by putting together a list of the most flattering comments I'd ever received for my writing and put it by my computer. It's still there, but I never think to look at it. It now strikes me as an act of hollow desperation like when in the movie *Cast Away*, Tom Hanks' character, stranded alone on a tropical island after his plane crashed, attempted to create a friend by drawing a face on a volley ball left from the wreckage, and naming it Wilson after the logo stitched into the label. What *did* help me to start writing on a more regular basis was a conversation I had one evening with a dear friend from graduate school. Cherise is a professor of literature at West Chester University and has always managed to write poetry, her true passion, along with her academic work. A few years ago, she and I had a long conversation in the car before saying goodbye after our annual summer get-together. She told me, 'You can always spare twenty minutes a day. Make this daily commitment. No matter how busy you are you can at least do that. Really, it's a gift that you give to yourself'. This advice was her gift to me. That, along with her willingness to read carefully anything I write at any stage, raw or finished, and to honestly share her responses and suggestions.

Of course, writing conferences, workshops and writers' retreats are held for exactly these communal benefits. But it took me awhile to realize their worth, probably because the few creative writing courses I took while at university engaged my spirit about as much as any church ever did. You knew that the other people in the class, a step away from strangers at best, read your work only because they were forced to. Nothing said this more than the sight of them one by one dumping the writer-of-the-day's pages into the trash as they left the classroom. Now, I regularly exchange work with a select group of people, Cherise among them, who are both friends and fellow writers. Unlike the workshops I encountered as a student, these groups developed organically out of shared affinities about life and writing, and most importantly a shared sense of writing as a sacred activity. Though I have a few trusted male readers on whom I rely for feedback on my work, it's probably no accident that these precious writing communities have been entirely made up of women. Meeting regularly, we create sacred spaces, of both withdraw and return, where our bodies are not only welcome, but often the source of inspiration. This is also probably why, though there have been men and women who have appreciated my work, it's been mostly women who've ever expressed *gratitude* for it. Better than praise, a reader's expression of gratitude is the best assurance of having returned from writing with something worthwhile to give back. I don't feel I could make the commitment to withdraw from the world into myself and my writing if I did not have this community to return to, a community built out of the faith we have in one another as people and as writers, and in writing as a general good.

I guess what I'm trying to say is that it takes a village, really a church, to raise a writer. After that, it's as natural as breathing.

Bibliography

Ansary, Tamim. 'Villages and Compounds'. *The New World Reader: Thinking and Writing about the Global Community*, edited by Gilbert H. Muller, 436–440. New York: Houghton Mifflin Company, 2005.

Freud, Sigmund. *Totem and Taboo*. Translated and edited by James Stratchey. New York: Norton, 1989.

Vincent, van Gogh. '371 To Theo van Gogh. The Hague, on or about Tuesday', 7 August 1883. *Vincent van Gogh the Letters, van Gogh Museum*. Viewed 6 June 2016. http://vangoghletters.org/vg/letters/let371/letter.html.

Oxford English Dictionaries. "Spirit". Accessed January 15, 2017. https://en.oxforddictionaries.com/definition/spirit.

CHAPTER 5

Expression of Humour in Persuasive Writing: Developmental Trends and Pedagogical Implication

Ekaterina Midgette and Sevil Nakisli

Abstract

There is a dearth of studies that explore the extent to which school-age students use conditional knowledge (i.e., using figurative language including humour) in argumentative writing for the purpose of audience engagement. Such knowledge is not only effective in developing the cognitive and linguistic skills that are essential for skilful argumentation, but also makes the very process of writing engaging and enjoyable. In this chapter, we analyse persuasive and argumentative essays of students ranging from 8 to 14 years of age for developmental trends (i.e., frequency of usage, identifiable patterns of usage, and effectiveness of usage, depending on age) in the ability to engage their readers through the use of humour and playful language. Our findings indicate that (1) children use unprompted humour more frequently than adolescents; (2) children tend to use good-natured forms of humour, whereas adolescents frequently resort to derogatory humour in their essays; (3) adolescents outperform children in the effectiveness of using humour for the purposes of persuasion. This study suggests that humour can indeed be used in critical thinking, which is the backbone of any persuasive writing.

Keywords

Writing instruction – humour – audience engagement – audience awareness – argumentative essays – persuasive essays

• • •

1 Introduction

In these times of the nationalization of the Language Arts curriculum in the U.S. and the exponential increase of teachers' accountability through high

 | DOI:10.1163/9789004394315_006

stakes testing both at the state and national levels, the learning environment of Reading and Writing blocks in elementary schools and English Language Arts classrooms at the middle school level has become more serious than ever. There is compelling evidence coming from the fields of neurology and educational psychology indicating that humour fosters retention of verbal information,[1] supports focus and attention during instructional tasks[2] and reduces anxiety associated with learning difficult material.[3] In spite of this, students are rarely, if ever, asked to write humorous texts or use elements of humour in composition.[4] There is a dearth of studies that explore the extent to which school-age students use conditional knowledge (i.e., using figurative language, including humour) in argumentative writing for the purpose of audience engagement. Such knowledge may not only be effective in developing the cognitive and linguistic skills that are essential for skilful argumentation,[5] but also make the very process of writing engaging and enjoyable. The purpose of this chapter is to analyse persuasive and argumentative essays of students ranging from 8 to 14 years of age (divided into two groups: Children and Adolescents) for developmental trends in the ability to engage their readers through the use of humour and playful language. Specifically, we were interested in exploring whether students were using humour more often or less often as they get older. Also, we wondered what kinds of humour are used by children versus adolescents and whether any of the humour categories overlap between the two age groups. In order to do that, we employed four categories of humour described by Ziff.[6] These categories are Broad Humour (i.e., a light-hearted joke), Wit (i.e., a sharp or sarcastic comment with the purpose of expressing an opinion), Dig (i.e., a derogatory remark with the intent of insulting the recipient) and Neutral Humour (i.e., word play for the sake of entertainment). Finally, we sought to investigate if students either age group were more effective in increasing the persuasive power of their essays through humour.

1 Tunku S. Badli, Zawyah Tunku and Mariam A. Dzulkifli, 'The Effect of Humour and Mood on Memory Recall', *Procedia – Social and Behavioural Sciences* 97 (2013): 252–257.

2 Pascal Vrticka, et al., 'Sex Differences during Humour Appreciation in Child-Sibling Pairs', *Society for Neuroscience* 8.4 (2013): 291–304.

3 Rod A. Martin, *The Psychology of Humour: An Integrative Approach* (Amsterdam: Elsevier Academic, 2007).

4 Kimberly H. Campbell, 'Eavesdropping on Contemporary Minds: Why We Need More Essays in Our High School Classrooms', *English Journal* 99.4 (2010): 50–54.

5 Edward S. Inch and Barbara Warnick, *Critical Thinking and Communication: The Use of Reason in Argument* (Boston, MA: Allyn and Bacon, 2002).

6 Selma S. Ziff, 'An Investigation of the Written Expression of Humour by Sixth-grade Gifted Children' (PhD diss., Virginia Polytechnic Institute and State University, 1982).

2 Using Humour to Persuade

Humour pervasively occupies human thought and written expression as a vehicle of conveying perceived incongruity in the immediate environment and our personal, social and political lives.[7] Humour operates by bisociation, which is defined as a creative way to connect self-consistent but habitually incompatible frames of reference,[8] essentially disrupting the rules of formal logic.[9] Questioning perceived absolutes, such as power and authority, broadening the possibilities for reasoning, challenging the dominant views and initiating public debate[10] are among the many functions of humour. Since Jonathan Swift's *Gulliver's Travels*,[11] humour (in this case social and political satire) has been explicitly connected with argumentative and persuasive writing.[12] Other authors point out that humour has a prominent role in questioning and negotiating reality (physical and existential) in religious texts of the Judaic-Christian tradition.[13] The playful approach that Abraham took when arguing his case to save Sodom and Gomorra[14] is one example of questioning the plan of the Almighty and engaging in negotiations with God.

In our times, commercial scripts are the epitome of using humour for the purpose of persuasion. In fact, one out of five advertisements make a persuasive appeal using incongruent and surprising messages with a strong humorous connotation.[15] Writers of commercial scrips intentionally and artfully use humour to persuade the audience. Humorous texts are a powerful tool to grab the attention of the audience (readership),[16] provided it is at an optimal level of complexity and novelty (i.e., it is not too hard and not too easy to understand

---

7 Giselinde Kuipers, 'The Sociology of Humour', *The Primer of Humour Research*, ed. Victor Raskin (Berlin: Walter De Grayter, 2008), 361–398.

8 Arthur Koestler, *The Act of Creation* (London: Hutchinson, 1964).

9 Aurelia Ana Vasile, 'The Illocutionary Side of Argumentation with Humour in Media Infotainment', *Journalism si Comunicare* 3.4 (2009): 47–53; Chris Smith and Ben Voth, 'The Role of Humour in Political Argument: How "Strategery" and "Lockboxes" Changed a Political Campaign', *Argumentation and Advocacy* 22 (2002): 110–128.

10 Don Waisanen, 'Jokes Inviting More than Laughter … Joan Rivers' Political-Rhetorical World View', *Comedy Studies* 2.2 (2011): 139–150; Smith, 'The Role of Humour', 110.

11 Jonathan Swift, *Gulliver's Travels* (Los Angeles: Millennium Publications, 2015).

12 Carol Reeves, 'Students as Satirists', *College Teaching* 44.1 (1996): 15–18.

13 Don Waisanen, Hershey H. Friedman and Linda Weiser Friedman, 'What's so Funny about Arguing with God? A Case for Playful Argumentation from Jewish Literature', *Argumentation* 29.1 (2015): 57–80.

14 Genesis 18:23–33 (New Revised Standard Version).

15 Fred K. Beard, 'One Hundred Years of Humour in American Advertising', *Journal of Macromarketing* 25.1 (2005): 54–65.

16 Martin Eisend, 'A Meta-Analysis of Humour in Advertising', *Journal of the Academy of Marketing Science* 37.2 (2009): 191–203.

the joke or that is not too surprising or too trite).[17] Consumers of the advertisement devote more cognitive resources to the message when they perceive a humorous incongruity and process a positive message about the benefits of the advertised product more favourably. In addition, not only does humour draw attention but it also makes the persuasive message more memorable, as it increases retention.[18] Of particular importance to argumentation, as opposed to the one-sided rhetoric of persuasion, humorous arguments have a quality of suppressing refutation, as they function as 'switches that would turn down the intensity of debate or refutation'.[19] This, in turn enhances a positive perception of the argument. In advertising, this effect is linked to what is known as a 'vampire effect':[20] using humour to distract the consumer from processing the benefits and negative attributes of the brand/company (i.e., a broader argument) by drawing the attention to a particular product (i.e., persuasive attempt).[21]

Humour permeates students' conversations in and outside of school, as children reflect on their lives. Encouraging students to use humour in their writing helps them to make a writing assignment more relevant and authentic to the purpose.[22] Humorous writing, particularly satire and parody, has also been used in the secondary classroom as a way to deconstruct the inequities that students perceive in their own lives.[23] Humour fosters creativity in the writing classroom, as humorous effect is often created by teaching students to argue their case in a unique writing format or change a perspective.[24] Using humour skilfully for the purpose of audience engagement also comes from creativity induced by language play, such as connecting incongruities through the use of surprising metaphors or hyperboles. It also requires attention to word choice, pace and sentence structure. In other words, as Bev Hogue[25] put it, 'good humour writing is good writing – period'.

---

17 Daniel E. Berlyne, *Conflict, Arousal, and Curiosity* (New York: McGraw Hill, 1960).

18 Martin Eisend, 'How Humour in Advertising Works: A Meta-Analytic Test of Alternative Models', *Marketing Letters* 22.2 (2010): 115–132.

19 Charles S. Gulas and Marc G. Weinberger, *Humour in Advertising: A Comprehensive Analysis* (Armonk, NY: M.E. Sharpe, 2006).

20 Robin B. Evans, *Production and Creativity in Advertising* (London: Pitman, 1988).

21 Martin Eisend, 'A Meta-Analysis of Gender Roles in Advertising', *Journal of the Academy of Marketing Science* 38.4 (2010): 418–440.

22 Nina Murakami, 'Not Just a Humorous Text: Humour as Text in the Writing Class', *Teaching English in the Two-Year College*, 34.1 (2006): 32–40.

23 Reeves, 'Students as Satirists', 15.

24 Bruce H. Weber, 'Closure in the Emergence and Evolution of Life: Multiple Discourses or One?' *Annals of the New York Academy of Sciences* 901.1 (2000): 132–138.

25 Bev Hogue, 'I'm Not Making This Up: Taking Humour Seriously in the Creative Nonfiction Classroom', *Pedagogy: Critical Approaches to Teaching Literature, Language, Composition and Culture* 11.1 (2010): 199–205.

We take the liberty to extend this statement by claiming that a good humorous argument is a good argument, perhaps not from the view of formal logic, but from the perspective of readership engagement, as witty, playful or sarcastic arguments inevitably and quickly get the reader to pay attention to the author's claim. In the virtual absence of empirical research on the role of humour in school-age children's persuasive and argumentative writing,[26] it is important to investigate whether or not and how school-age children use humour in argumentative and persuasive written discourse in the absence of prompting or any other form of intervention targeted at eliciting humour in writing. Such investigation contributes to current research in compositional pedagogy by outlining the strengths and needs that students demonstrate in their attempts to engage their audience through humour. The outcomes of this study can also be used as a base for exploration of effective methods in teaching students to increase the persuasive quality of their essays with humorous argumentation.

3 Method

Participants and Data Collection: Written essays were collected from 455 female and male participants ranging from 8 to 14 years of age. The essays written by students with special needs, including gifted learners, and by English Language Learners were excluded from the analysis. The number of participants in the Children group was 251, whereas the number of participants in the Adolescent group was 204. Given the uneven number of the participants in the two group, the findings of the study were reported and compared in percentages.

Two different prompts were used in the study. Third graders and fourth graders were given an assignment to write a letter to parents or classroom teachers asking for a privilege. Seventh and eighth graders were asked to write an essay expressing their opinion on whether or not they would want to become a celebrity and give reasons for their hypothetical choice. The prompts were designed to be of interest to the corresponding age group and unbiased in terms of culture and gender. The students were given instructions to plan, draft and revise their essays according to the practices of good writing they learned in school. The participants were not aware of the purpose of the study.

---

26 To our knowledge and confirmed by other researchers, such as Vasile, 'The Illocutionary Side of Argumentation', 51; Hogue, 'I'm Not Making this Up', 200.

Research Questions: The research questions that guided this investigation were:

Are there identifiable patterns in frequency and types of humour used by children (age 8 to 10) versus adolescents (age 12 to 14) in written essays?

Are there identifiable patterns in the effectiveness of humour usage to support the overall argument in written essays of children (age 8 to 10) versus adolescents (age 12 to 14)?

Analysis: Three individuals worked independently to identify incidents of humour in students' essays. They were classified according to technique used and the category to which every technique belongs. The categories emerged from Ziff's study[27] who investigated the expression of humour in written work of gifted sixth-graders. Ziff[28] identified 4 categories of humour technique: Broad Humour, Wit, Dig and Neutral Humour. Wit is distinguished from Broad Humour as being higher on the cognitive scale and where Dig is lower than Wit on the humanistic scale. Broad Humour is defined as good natured leisurely humour based on reality or fantasy and geared towards a wide audience, whereas Wit is persuasive, sharp, and tends to reinforce a pre-existing attitude and has a limited audience. Dig shares these characteristics with Wit but is more hostile. Neutral does not have a persuasive power, it is best characterized as linguistic play without a purpose other than to entertain the reader and have fun with words.

Individual incidents of humour were rated on the scale from 1 to 3. A score of 3 indicates that the incident of humour is explicitly connected with the central claim of the argument and is used to persuade the reader; a score of 2 indicates that humour is used in support of one of the reasons given by the author's claim but not linked to the central claim directly. A score of 1 was assigned to indicate that the incident is a use of playful language that aids in audience engagement but is not connected to the overall argument of the essay or any of the supporting reasons.

The following statistics were obtained to explore the research questions:

1. Frequency of humour use by gender, grade and group (i.e., children versus adolescents).
2. Frequency of humour techniques and categories by gender, grade and group.
3. Frequency of humour factor scales by gender, grade and group.

---

27 Ziff, 'An Investigation of the Written Expression of Humour by Sixth-grade Gifted Children', appendix.

28 Ibid.

4 Results

Finding 1: The analysis of 455 essays revealed that 119 students used humour in their essay, 82 (69%) of them were children and 37 (31%) were adolescents. In Children, 44 (37%) of humour users were girls and 38 (32%) were boys. In Adolescents, 22 (18%) of humour users were boys and 15(13%) were girls. Analysis was based on the 175 humour instances found in the essays of the 119 students. Out of the 175, 125 (71%) were used by children, whereas 50 (29%) were used by adolescents (Appendix 1).

It is possible to interpret these findings from two, not necessarily mutually exclusive, perspectives. First, the culture of adolescent writing in middle school in the U.S. is dominated by the strict standards of the state curriculum and leaves little room for playful language and other forms of linguistic experimentation and creativity. In New York State, for example, the seventh grade unit that is designed to teach reading comprehension strategies and to help students develop their written argumentation skills in response to text is based on the transcript of Cesar Chavez's speech to the Commonwealth Club in 1984 advocating for unionization of farm workers in southern California.[29] Truly, it is a worthy educational endeavour to inform students in New York of the plight of farm workers in southern California 25 years ago. This text, however, does not elicit the same kind of opinions as texts adolescents consider to be their own. The emphasis of the curriculum is on their ability to demonstrate declarative knowledge of structural elements of the argumentative genre (i.e., inclusion of position, three reasons, rebuttal, etc.). In contrast, students at the elementary level in the U.S. operate in a more flexible writing environment. Children are not assigned letter or percentage grades for their written work and are commonly encouraged to have fun with writing and discuss it with peers and the teacher. Persuasive writing in particular is often taught using humorous examples, such as 'I Wanna Iguana'[30] or 'The True Story of Three Little Pigs'[31] and other fractured fairy-tales.

The topic and content of the assignment could explain the reluctance of adolescents to use humour in their writing. Children in the present study were prompted to write to familiar people (i.e., parents or teachers), whereas the audience assigned to adolescents was more distant and vague (i.e., writing to

29 New York State Education Department, *New York State ELA Curriculum Grade 7*, viewed 19 January 2017, https://www.engageny.org/common-core-curriculum.

30 Karen Kaufman Orloff and David Catrow, *I Wanna Iguana* (New York: Putnam, 2004).

31 Jon Scieszka and Lane Smith, *The True Story of the Three Little Pigs* (New York: Viking Kestrel, 1989).

voice their opinion). Perhaps perceiving a concrete individual as audience, while writing, may evoke the use of internal dialogue with the posited audience. Students routinely use humour in and outside of school. Having an assignment that mimics authentic interaction might help make a connection between their genuine style of communication and the class assignment. Further research is needed to investigate the patterns of humour use by children versus adolescents in writing on topics of varying degrees of intimacy to confirm or exclude the developmental factor in connecting with the audience through humour.

Finding 2: Patterns emerged when analysing humour usage by category. Children clearly gave preference to Broad Humour, as 69% of all children who used humour in their writing used a good-natured joke type of humour, exaggeration (33%) and absurdity (8%) being the most frequently used techniques within this category. Another popular technique, repetition (Neutral Humour category), was used by 14 (12%) students.

In adolescent group, the only type of technique used in the Broad Humour category was exaggeration. 4 (4%) boys and 1 (1%) girl used this technique. Insult, mockery and stereotyping, on the other hand, oozed from adolescents' essays, making Dig the highest ranking category for this group of students. The humour techniques of sarcasm in Wit category and insult in Dig category were used by the most number of adolescents: sarcasm was used by 4 boys and 3 girls and insult was used by 4 girls and 3 boys (Appendix 2).

These data show us that Dig was non-existent in the work of children, and Neutral Humour was not a part of adolescents' writing. Wit in children's writing was found mostly in the visual representation of humour (i.e., cartoons), whereas adolescents used sarcasm, irony and satire but to a lesser extent than other forms of humour. Boys and girls in both groups did not significantly differ in their preference of technique. The exceptions being that girls in the Children group were more likely than boys to make a joke by drawing cartoons in their letters and boys were more actively engaged in exaggeration than girls.

Although it might be tempting to interpret these results from the developmental point of view (i.e., adolescents falling short of humanistic considerations for others when trying to be funny), we give adolescents the benefit of a doubt by taking into account the difference between the prompts assigned to the two groups. Children might not have been in the position to look down upon their audience when writing with the purpose of eliciting a favour, a present or permission from their parents and teachers. The topic given to the adolescent group, on the other hand, could have provoked a judgmental attitude on the part of the students, hence the choice of using a more derogatory form of humour. A future study should examine students' responses to the same topic in order to avoid such ambiguity in interpretation of the results.

Finding 3. Scoring incidences of humour with the Humour Factor scales revealed that adolescents outperformed children in the effectiveness of using humour for the purposes of persuasion (Appendix 3).

The results suggest that the students in fact enhance the persuasiveness of their appeal by using humour, albeit not to the fullest extent. Among 119 humour users, only 8 students used humour which are directly related to the argument (scale 3), 6 of them were adolescents and only 2 were children. Seven out of eight students who used humour to support the central claim of the arguments (scale 3) were boys whereas only 1 was a girl. This result shows that teens, especially boys, demonstrated a better ability in connecting a joke with the central claim of their essays and make the overall argument appear more convincing through the 'vampire effect'.

Nine (36%) adolescents used humour to illustrate a supporting reason (scale 2). Children also frequently employed humour to illustrate a supporting reason, though lightening up the essay and playing with language (scale 1) appeared to be the predominant function of humorous instances found in their essays (i.e., 47 (80%) children verses 12 (20%) of adolescents).

5 Discussion

The findings of this study suggest that teachers at all grade levels should create a classroom atmosphere that encourages humour usage through humorous examples of persuasive and argumentative texts that are abundant in our politicized and criticism-driven culture. Adolescents in particular may find the comic release of their emotions particularly effective. Guidance in the use of humour to evaluate social problems can be an effective way to promote critical thinking and prepare students for becoming a part of adult society where humour is a highly valued social skill.

If these results are confirmed by future research that explores adolescents' response to a range of topics, it might prove important to guide adolescent writers in redirecting their frustrations from the biting hostility of Dig into witty satire, which builds on the exaggeration frequently used by teens, with the purpose of unveiling and making sense of the follies and inequities in the world around them.[32] Social satire is a particularly relevant humour technique to use in argumentative writing. Putting on satirist's hat requires a firm view of one's own position regarding the issue under attack and having a thorough

32 Reeves, 'Students as Satirists', 15.

factual knowledge of the issue one is satirizing and both of these are essential elements of an effective argumentative essay.

The appearances of humour usage supporting the overall argument of the essays or one of the reasons indicate that such purposeful use of humour to strengthen the persuasive appeal of the essays is within reach for students, even for young writers. Future studies should be directed to the examination of the effects of explicit instruction in using humour for the purposes of persuasion on the holistic persuasiveness of student essays. Such interventions would include analysis of the types of humour that are developmentally appropriate for a given group of students using examples from editorials, articles, blogs and children or young adult literature. This should be followed by an active demonstration and guided practice in using humour to support the overall argument and supporting reasons, and giving students a choice of topics that are likely to elicit a humorous response (e.g., 'If I were the school principal').

Relationships indicated in this study generate numerous avenues for instructional interventions that could help us understand the ways in which school-age students use humour in written expression to reflect on the world around and within them. This study suggests that humour can indeed be used in critical thinking, which is the backbone of any persuasive writing. Teachers of writing need to capitalize on a child's ability to use humour to question, reflect and release emotions. Teaching how to write humorously through a playful change of perspectives, unexpected associations, change of style, humorous topics, unique writing formats, analysing and mimicking humorous texts will create a relaxing and enjoyable writing environment conducive to creativity.

Bibliography

Badli, Tunku S., Zawyah Tunku and Mariam A. Dzulkifli. 'The Effect of Humour and Mood on Memory Recall'. *Procedia – Social and Behavioral Sciences* 97 (2013): 252–257.

Beard, Fred K. 'One Hundred Years of Humour in American Advertising'. *Journal of Macromarketing* 25.1 (2005): 54–65.

Berlyne, Daniel E. *Conflict, Arousal, and Curiosity*. New York: Mc Graw Hill, 1960.

Campbell, Kimberly H. 'Eavesdropping on Contemporary Minds: Why We Need More Essays in Our High School Classrooms'. *English Journal* 99.4 (2010): 50–54.

Eisend, Martin. 'A Meta-Analysis of Humour in Advertising'. *Journal of the Academy of Marketing Science* 37.2 (2009): 191–203.

Eisend, Martin. 'How Humour in Advertising Works: A Meta-Analytic Test of Alternative Models'. *Marketing Letters* 22.2 (2010): 115–132.

Eisend, Martin. 'A Meta-Analysis of Gender Roles in Advertising'. *Journal of the Academy of Marketing Science* 38.4 (2010): 418–440.

Evans, Robin B. *Production and Creativity in Advertising*. London: Pitman, 1988.

Gulas, Charles S. and Marc G. Weinberger. *Humour in Advertising: A Comprehensive Analysis*. Armonk, NY: M.E. Sharpe, 2006.

Hogue, Bev. 'I'm Not Making this Up: Taking Humour Seriously in the Creative Nonfiction Classroom'. *Pedagogy: Critical Approaches to Teaching Literature, Language, Composition, and Culture* 11.1 (2010): 199–205.

Inch, Edward S. and Barbara Warnick. *Critical Thinking and Communication: The Use of Reason in Argument*. Boston, MA: Allyn and Bacon, 2002.

Koestler, Arthur. *The Act of Creation*. London: Hutchinson, 1964.

Kuipers, Giselinde. 'The Sociology of Humour'. *The Primer of Humour Research,* edited by Victor Raskin, 361–398. Berlin: Walter De Grayter, 2008.

Martin, Rod A. *The Psychology of Humour: An Integrative Approach*. Amsterdam: Elsevier Academic, 2007.

Murakami, Nina. 'Not Just a Humorous Text: Humour as Text in the Writing Class'. *Teaching English in the Two-Year College*, 34.1 (2006): 32–40.

Orloff, Karen K., and David Catrow. *I Wanna Iguana*. New York: Putnam, 2004.

New York State Education Department, *New York State ELA Curriculum Grade 7*, viewed 19 January2017, https://www.engageny.org/common-core-curriculum.

Reeves, Carol. 'Students as Satirists'. *College Teaching* 44.1 (1996): 15–18.

Scieszka, Jon and Lane Smith. *The True Story of the Three Little Pigs*. New York: Viking Kestrel, 1989.

Smith, Chris and Ben Voth. 'The Role of Humour in Political Argument: How "Strategery" and "Lockboxes" Changed a Political Campaign'. *Argumentation and Advocacy* 22 (2002): 110–128.

Vasile, Aurelia A. 'The Illocutionary Side of Argumentation with Humour in Media Infotainment'. *Journalism si Comunicare* 3.4 (2009): 47–53.

Vrticka, Pascal, Michelle Neely, Elizabeth Walter Shelly, Jessica M. Black, and Allan L. Reiss. 'Sex Differences during Humour Appreciation in Child-Sibling Pairs'. *Society for Neuroscience* 8.4 (2013): 291–304.

Waisanen, Don. 'Jokes Inviting More than Laughter ... Joan Rivers' Political-Rhetorical World View'. *Comedy Studies* 2.2 (2011): 139–150.

Waisanen, Don, Hershey H. Friedman, and Linda Weiser Friedman. 'What's so Funny about Arguing with God? A Case for Playful Argumentation from Jewish Literature'. *Argumentation* 29.1 (2015): 57–80.

Weber, Bruce H. 'Closure in the Emergence and Evolution of Life: Multiple Discourses or One?' *Annals of the New York Academy of Sciences* 901.1 (2000): 132–138.

Ziff, Selma S. 'An Investigation of the Written Expression of Humour by Sixth-grade Gifted Children'. PhD dissertation, Virginia Polytechnic Institute and State University, 1982.

Appendix 1

Demographic composition and humour usage of the student

Demographics	Students		Humor users		Humor usage	
	(#)	(%)	(#)	(%)	(#)	(%)
Children (Age: 8–10)	251	55.2	82	68.9	125	71.4
3rd Grade (Age: 8 and 9)						
Boys	55	12.1	18	15.1	19	10.9
Girls	70	15.4	19	16.0	31	17.7
Total	125	27.5	37	31.1	50	28.6
4th Grade (Age: 9 and 10)						
Boys	68	14.9	20	16.8	29	16.6
Girls	58	12.7	25	21.0	46	26.3
Total	126	27.7	45	37.8	75	42.9
Gender (Age: 8–10)						
Boys	123	27.0	38	31.9	48	27.4
Girls	128	28.1	44	37.0	77	44.0
Adolescents (Age: 12–14)	204	44.8	37	31.1	50	28.6
7th Grade (Age: 12 and 13)						
Boys	57	12.5	11	9.2	12	6.9
Girls	55	12.1	6	5.0	9	5.1
Total	112	24.6	17	14.3	21	12.0
8th Grade (Age: 13 and 14)						
Boys	49	10.8	11	9.2	16	9.1
Girls	43	9.5	9	7.6	13	7.4
Total	92	20.2	20	16.8	29	16.6
Gender (12–14)						
Boys	106	23.3	22	18.5	28	16.0
Girls	98	21.5	15	12.6	22	12.6
Total (Age: 8–14)	455	100	119	100	175	100

Appendix 2

Humour notation and categories

		Number of students						Number of humour used by					
		Children			Adolescent			Children			Adolesc.		
		Boys	Girls	All	Boys	Girls	All	Boys	Girls	All	Boys	Girls	All
Broad													
	Absurdity	3	6	9	0	0	0	3	12	15	0	0	0
	Exaggeration	22	17	39	4	1	5	25	29	54	4	1	5
	Self-Derision	0	2	2	0	0	0	0	3	3	0	0	0
	Taboo Reference	1	1	2	0	0	0	2	1	3	0	0	0
	Total	26	26	52	4	1	5	30	45	75	4	1	5
Wit													
	Cartoon	2	7	9	0	0	0	2	10	12	0	0	0
	Facetiousness	0	1	1	0	0	0	0	2	2	0	0	0
	Irony	1	2	3	3	1	4	4	5	9	3	1	4
	Mock Serious	1	0	1	0	0	0	1	0	1	0	0	0
	Sarcasm	0	0	0	4	3	7	0	0	0	8	3	11
	Satire	0	0	0	3	0	3	0	0	0	4	0	4
	Total	4	10	14	10	4	14	7	17	24	15	4	19
Dig													
	Mockery	0	0	0	3	1	4	0	0	0	0	8	8
	Insult	0	0	0	3	4	7	0	0	0	0	9	9
	Stereotype	0	0	0	2	3	5	0	0	0	0	7	7
	Total	0	0	0	8	8	16	0	0	0	0	24	24
Neutral													
	Repetition	7	7	14	0	1	1	10	14	24	0	1	1
	Slang	1	1	2	0	1	1	1	1	2	0	1	1
	Total	8	8	16	0	2	2	11	15	26	0	2	2
General	Total	38	44	82	22	15	37	48	77	12	19	31	50

Appendix 3

Humour notation and categories

	Scale 3		Scale 2		Scale 1	
	Humor	% in Total Humor	Humor	% in Total Humor	Humor	% in Total Humor
Children (Age: 8–10)						
3rd Grade (Age: 8 and 9)						
Boys	0	0	6	12	12	20
Girls	0	0	4	8	15	25
Total	0	0	10	19	27	46
4th Grade (Age: 9 and 10)						
Boys	2	25	7	13	11	19
Girls	0	0	16	31	9	15
Total	2	25	23	44	20	34
Gender (Age: 8–10)						
Boys	2	25	13	25	23	39
Girls	0	0	20	38	24	41
Total Children	2	25	33	63	47	80
Adolescents (Age: 12–14)						
7th Grade (Age: 12 and 13)						
Boys	1	13	4	8	6	10
Girls	0	0	3	6	3	5
Total	1	13	7	13	9	15
8th Grade (Age: 13 and 14)						
Boys	4	50	5	10	2	3
Girls	1	13	7	13	1	2
Total	5	63	12	23	3	5
Gender (12–14)						
Boys	5	63	9	17	8	14
Girls	1	13	10	19	4	7
Total Adolescent	6	75	19	37	12	20
Total (Age: 8–14)	8	100	52	100	59	100
% of Scales		7		44		50

PART 2

Meeting the Master

CHAPTER 6

Gandhi, the Journalist: When, Where, How, and What He Wrote

David W. Bulla

Abstract

This chapter explores Mohandas K. Gandhi as a writer—where he wrote (often at home, usually in a communal, ashram atmosphere), when he wrote, what he wrote about, how he saw writing, what it meant to his life, particularly the place it had both in his spiritual quests, as well as being a practical tool for advancing his political and religious causes. Indeed, Gandhi was a journalist for forty-four years. He edited seven newspapers during that span, starting with *Indian Opinion* in 1903 when he was in South Africa and with *Young India* when he returned to his native land in 1915. These enterprises were family-oriented, and the father served as editor, though never officially as publisher (and his newspapers were never copyrighted). For Gandhi, writing was central to not just his causes but to his entire being. Writing was not only a habit of being; it was also Gandhi's way of expressing himself to the masses, especially since he was never fond of his oratory skills. As an attorney, he felt more comfortable with the precision of writing, and his roots in writing go back to the renaissance in printing that occurred on the West Coast of India in the late nineteenth century when publishing soared on the subcontinent and the young Gandhi feasted on the available literature before going off to London to study law. We can understand how important writing was to Gandhi by taking a brief audit of his material possessions at his final residence, Birla House, in New Delhi. In his Birla bedroom, one sees a mattress, wire-framed eyeglasses, a glasses case, eating utensils, a walking stick, and a small writing desk. Walking, spinning, praying, and writing were the essential activities of his ritualistic life.

Keywords

England – Gandhi – India – journalism – pamphlet – *Satyagraha* – South Africa

•••

 | DOI:10.1163/9789004394315_007

1 Writing Roots of a Journalist

Mohandas K. Gandhi, who practiced journalism for more than four decades, had no pretence to become a novelist, playwright, or poet. Rather, writing for Gandhi was a practical matter and it served his philosophy, *satyagraha*. For the Indian, putting pen to paper—and publishing his newspapers and books—was a nonviolent political act. Like burning a pass card in a cauldron, refusing to sit in third class on a train, refusing to be thumb-printed for registration, or choosing to go to jail rather than paying a tax, Gandhi used writing as a way to protest the injustices he encountered in South Africa and India. Indeed, writing was a way of convincing people that the evidence he had accumulated told a story, and that those people not only needed to be aware of the evidence, but they ought to act upon it. The Mahatma used writing as a substitute for courtroom argumentation. Gandhi saw writing not only as a way to persuade people to his perspective, but it was also part of his being—which always came back to his search for truth. As much as it was a technological tool for communicating his thoughts, writing also was an explorative apparatus that helped him understand his world and himself by writing about education, equality, independence, nonviolence, religion, and vegetarianism.

Reading is the constant companion of writing, and Gandhi's childhood reading included the *Ramayana,* Gujarati drama, and the *Manusmriti*, an ancient legal text.[1] Of those, the *Ramayana* probably produced the greatest impact, as he began to read it at the suggestion of a maid who thought it would help him overcome his childhood fear of ghosts and criminals. For the rest of his life, he would repeat verses from the *Ramayana* when he was despairing.[2] In 1888, at age nineteen, he landed in London, where he studied law. Gandhi's law reading included a Roman law textbook (in Latin), Herbert Broom's *Commentaries on the Common Law*, and William Douglas Edwards' *The Law of Property in Land.* Gandhi also read two British newspapers, the *Daily Telegraph* and the *Pall Mall Gazette*.

Gandhi also began to read the Edwin Arnold English translation of the *Bhagavad Gita*. Gandhi stated that the emphasis in the *Gita* on being dispassionate and serving a higher purpose made a great impression on him.[3] From the *Gita*, he also took away the idea of maintaining an active life as a form of

1 Bhanbani Bhattacharya, *Gandhi the Writer* (New Delhi, India: National Book Trust, 2002), 8–9.

2 Ibid., 9.

3 Ibid., 13–14.

worship itself.[4] Two important books he read in his London years were Henry Salt's *Plea for Vegetarianism* and Howard Williams' *Ethics of Diet*. Salt's book he discovered at a vegetarian restaurant in London, and it reinforced his mother Putlibai Gandhi's view that only a vegetarian could be healthy and pure.[5] Williams held that the great prophets through the years had all been vegetarians, and Gandhi was moved by these examples.[6] At this time, he joined the English Vegetarian Society and in 1889 began to read its London-based journal, *The Vegetarian*. Two years later, Gandhi began writing for *The Vegetarian*.

2 In London

Gandhi's first article in *The Vegetarian* concerned vegetarianism in India. It was the first of a five-instalment series on the subject. That initial article discusses how only certain castes in India have to be vegetarian, and that vegetarians and meat-eaters are divided on religious grounds. Due to India's extreme poverty, Gandhi wrote that most Indians could not afford or even find meat if they were disposed to eat animal flesh. He also commented that Indian vegetarians did not eat eggs, for 'to eat an egg is equivalent to killing life', he noted.[7] Also, at this time, he began to write against inebriation. He certainly felt that alcohol intake was poor for one's health and contradictory to the kind of diet he followed. In the second article, Gandhi described typical Indian meals. Gandhi wrote that for most Indians, due to the heat, there were only two meals per day, one at 10 a.m. and the other at 8 p.m. He remarked that the morning meal consisted of bread ('cakes'), rice, vegetables, and milk. The evening meal would be similar, 'but the quantity is less and the vegetables fewer at this meal'.[8] Sweets are not included, except for the upper classes. In the third instalment, he described Indian bread as being like wheat cakes. He concluded:

> What meat is to the ordinary Englishman, cake is to the Indian, be he a Vegetarian or a meat-eater, for in India a meat-eater does not in the writer's opinion regard his meat as an absolute necessity, but takes it rather as a side dish to help him, so to speak, in eating the cakes.[9]

4 Arthur Herman, *Gandhi & Churchill: The Epic Rivalry That Destroyed an Empire and Forged Our Age* (London, UK: Arrow Books, 2009), 79.
5 Bhattacharya, 12.
6 Ibid., 12.
7 Mohandas K. Gandhi, 'Indian Vegetarians', *The Vegetarian*, London, UK, 7 February 1891.
8 Mohandas K. Gandhi, 'Indian Vegetarians', *The Vegetarian*, London, UK, 14 February 1891.
9 Mohandas K. Gandhi, 'Indian Vegetarians', *The Vegetarian*, London, UK, 21 February 1891.

In one of his last articles in *The Vegetarian*, the Indian writer moved into the overtly political realm. Titled 'English Misrule in India', Gandhi's piece praises the British government for being 'one of the most civilised in the world' and does not tolerate injustice, but then he gets to his main point: a new rule has been put in the place that all attorneys currently practicing law in Gandhi's home district in Gujarat (Kathiawar) have had their licenses revoked. They could only regain their right to practice law by paying an annual fee of thirty rupees—which the law student in London found unfair. Gandhi argued in the article:

> It seems that the notification indirectly aims at destroying the independence of every member of the Bar, for it becomes entirely impracticable for the legal practitioners to go on with their work, as they are meant to be at the mercy of the Political Agent (British governor of the area) for renewing their licenses every year. In no civilised government the independence of lawyers is not respected. When the freedom of press and freedom of debate are the true birth rights and privileges of the English people, why should not a small portion of them be liberally extended to Indians who are really in need of it and who have every right to claim a fair share of them from the British Government?[10]

This would become the template for Gandhi's future persuasive writing: Praise the British political system while detailing contradictions that unfairly impacted some of her subjects. Gandhi's writing for *The Vegetarian* was his first foray into mass communication, and it was starting to evolve toward something different. His writing was making the transition to a new topic—what today would be called human rights—and the protection of those rights for minorities and the dispossessed. Vegetarianism itself was seen as unconventional in England at the time Gandhi studied law, and he received pressure from some of his fellow students to abandon the practice.[11]

According to one biographer, Gandhi's prose at this time was 'frank, unpretentious, and rich in detailed information'.[12] That would become the template for Gandhi's verbal style for his entire career as a journalist: an emphasis on clearly, simply written detail to convince the reader of his main point. 'His sentences were simple and lucid', wrote biographer S.N. Bhattacharyya.[12] Gandhi did not show off verbally, and his newspapers did not have space for verbosity.

10 Mohandas K. Gandhi, 'English Misrule in India', *The Vegetarian*, London, UK, 29 July 1893.

11 Bhattacharya, 12.

12 Ibid., 17.

3 South African Writer

After his London years and a brief return to India, Gandhi was off to South Africa, where he began to write for public consumption in earnest, as he grew fond of voicing his opinion of matters of public opinion. First there were letters to the editors of newspapers in Durban. Next, he published pamphlets. The first pamphlets came in 1895 with these titles: (1) *An Appeal to Every Briton in South Africa* and (2) *The Indian Franchise*. These were followed by his first mass publication, the 1896 pamphlet titled *The Grievances of the British Indians in South Africa*. Gandhi published this so-called *Green Pamphlet* (because of the cover colour) in India, and the first edition went to ten thousand. It detailed incidents of mistreatment of Indians by white South Africans. Gandhi toured India promoting it, giving speeches to the public and interviews to newspapers. In the *Green Pamphlet*, he describes the symptoms of mistreatment and the effect such mistreatment had on Indians both legally and psychologically. Gandhi writes the policies, laws, and customs of racial hatred created an atmosphere of 'deep-seated hatred towards the Indians' that was 'reproduced all over South Africa'.[13]

His granddaughter Ela Gandhi, former member of the South African Parliament and a resident of Durban, said that her grandfather focused on his audience when he determined what was worth writing about and how he should write about it:

> He always said we should keep in mind what effect it [each article] would have on a person, on society when we determined what would go into the paper—into our paper—and what is the kind of news we do put into our paper; what did you want to promote. Don't just write. We have to think about the consequences of that particular sentence. An article should have a positive outcome for the reader. If it doesn't, then that story is not worth telling.[14]

In other words, her grandfather saw the words he wrote as being educational, as being uplifting, and even as being inspirational. The purpose of his journalistic writing was to improve the community Gandhi served.

---

13 Mohandas K. Gandhi, 'The Grievances of the British Indians in South Africa', self-published, Madras, India, August 14, 1896, 5. Hereafter, 'The Green Pamphlet'.

14 Ela Gandhi, interview by David W. Bulla, Durban, South Africa, 2 July 2014. Ela Gandhi's father, Manilal Gandhi, would edit and publish *Indian Opinion* for several decades after his father returned to India.

At a high school in Madras, in 1896, the young lawyer from South Africa told his audience that his main goal with the *Green Pamphlet* was to educate Indians about the conditions their countrymen faced in South Africa. He also wanted to reach the English public, to seek their support in getting the authorities to address Indian grievances. Gandhi reported the *Times of London* had written eight articles about the pamphlet. 'That alone has raised us a step higher in the estimation of the Europeans in South Africa', he wrote.[15]

The relationship to his audience was essential in his thinking about how writing functioned. He almost certainly learned this from his days in London as a law student preparing to argue in open court. Gandhi wanted to persuade his readers to his viewpoint, but who was his audience? The simple answer was Indians. In 1903, he entered journalism in earnest with *Indian Opinion*, in Durban. As its leading voice, Gandhi would only consider issues 'that directly affect Indians in this country' and 'would advocate the cause' of the Indian community in South Africa. The newspaper would contain the work of 'competent writers', both European and Indian.[16] When he began editing the paper, *Indian Opinion* only had 687 readers. Gandhi would more than double the readership, but it never was going to be very commercial in its orientation. Indeed, not very many South African Indians were literate, even in their native languages.[17] He envisioned most of his readers at this time to be the merchant class. *Indian Opinion* would take up issues such as unfair taxes, the pass laws, and the plight of Indian indentured servants in South Africa. Yet the Indians of South Africa were divided by caste, by religion, and by language. When his *satyagraha* campaigns began, Gandhi would begin to lose his merchant readers, who could not afford to adhere to his nonviolent direct action ethos because it would run them out of business. He also printed *Indian Opinion* initially in English, Gujarati, Hindi, and Tamil. He would drop the Hindi and Tamil versions because of the practical issues involving having so many different kinds of types to set at his Durban newspaper printing operation. That left the English and Gujarati sections of the weekly *Indian Opinion*, and over time a very different editor emerged in each of those two sections of the paper. South African literature professor Isabel Hofmeyr points out that in the English section of the Durban newspaper, the editor signed off as M.K. Gandhi, and in the Gujarati he was Mohandas Karamchand Gandhi. In the former, the tone of writing is 'polite'

15 Mohandas K. Gandhi, *The Collected Works of Mahatma Gandhi, Volume Two, 1896–1897* (New Delhi, India: Government of India, Publications Division, 1959), 155.

16 'The "Indian Opinion": The Policy', *Indian Opinion*, 4 June 1903, 2.

17 Nagindas Sanghavi, *The Agony of Arrival, Gandhi: The South African Years* (New Delhi, India: Rupa Publications, 2006), 263.

and formal, but in the latter he is 'outspoken, sometimes hectoring, emotional, passionate, truth-telling and personal'.[18] Furthermore, with the Gujarati reader, Hofmeyr comments, he is much more instructional.

4 Home to India

When he went back to India in 1914, he would start six more newspapers, including *Young India, Harijan, Harijan Bandu, Harijan Sevak, Navajivan,* and *Satyagrahi.* All would have informative and persuasive functions. Thus, the style of writing, observes Gandhi biographer Bhabani Bhattacharya, was 'simple, straightforward, orderly, precise; well-reasoned, lucid, adequately condensed; and above all else, single-minded, persuasive'.[19] All of the newspapers delved into the issues Gandhi felt passionate about. Indeed, they were advocacy journals, espousing various causes, and, of course, the greatest cause would be Indian independence from Great Britain. These were never mainstream newspapers 'selling news', so to speak, for mainly commercial reasons. By this time, his audience would be much wider than the one he had in South Africa. It would be those Indians who failed to benefit from British rule, and in this estimation, his imagined audience would have been millions of Indians—although clearly none of his newspapers ever had circulations of more than forty-five thousand, with the South African newspaper *Indian Opinion* only having three thousand at its height.

Gandhi slept very little, and in terms of his routine as a writer, Gandhi usually rose at 3 of 4 o'clock each morning. Probably his restless mind was most alert then, and his hand most rested. The writer would start his day with work on articles he was writing for one of his newspapers or correspondence to fellow letter writers around the world. He usually wrote with fountain pen and paper—not a typewriter and the personal computer would not be invented for decades after his death—and he sat with his knees folding under him with the paper on top of his legs. He does not look comfortable, and one wonders how many writers in the twenty-first century could fold their legs under them while sitting on the floor and still feel like writing. Gandhi wore his eyeglasses and put pen to paper with the right thumb and index finger doing all the action while the side of that right hand rests on the paper. His lips look like they are pouting, and the journalist writes with an intense look on his face. Gandhi

18 Isabel Hofmeyr, *Gandhi's Printing Press: Experiments in Slow Reading* (Cambridge, MA: Harvard University Press, 2013), 80.

19 Bhattacharya, *Gandhi the Writer*, 112.

tended to write in the same place he slept—for example, when he lived with Herman Kellanbach in Johannesburg, Gandhi wrote in his bed in a loft that he reached by a steep, narrow ladder. He wrote in a similar place in his ashram in Gujarat and the same at Birla House in Delhi, his last residence before his assassination. You see a squat stand next to his mattress, so that he can quickly move from sleep to writing. You see him writing while he is talking to dignitaries, such as his colleague Jawaharal Nehru. In one photo taken of Gandhi and Nehru, you see the prime minister has to literally come down to Gandhi's level, which mean's sitting on the floor very close to the Mahatma. Listening to independent India's first prime minister, Gandhi is still writing.

In the photographs of Gandhi writing, we usually see fountain pens and paper, but also newspapers or newspaper clippings. Most of the time, the setting is on a floor or a bed (Indian beds are often just mats on a floor). The rooms where his writing occurs are spare. The walls are unadorned. The room at Birla House where he spends the last days of his life is white with a floor bed. The rooms seem almost empty. Everything bespeaks a monastic life, and that too is the theme of Gandhi as a writer. The spaces where he wrote were purely minimalistic, and his worldview seems so horizontal since where he reads and writes is not standing up, but lying prostrate. Writing lying down somehow seems so passive, yet Gandhi's nonviolent dissent is anything but passive. It was confrontational.

His handwriting, in English, is jagged, as if there is tension. Maybe it is rushed. Perhaps he is in a hurry to get from one task to another. Gandhi is not just a writer, but he feels that recording, chronicling, arguing must be done. In order to get down all that he wants to record requires him to write without a great deal of attentiveness to good penmanship.

5 Conclusion: Why Gandhi Wrote

Why did Mohandas K. Gandhi write? The best answer is to spread his ideas, although most writers write for ego, and one suspects that this was the case for Gandhi too. Still, the more accurate answer is to search for truths. He said: 'I write to propagate my ideas' and 'to peep into myself and make discoveries of my weaknesses'. In another sense, Gandhi saw his writing as a kind of partnership between writer and reader, and the two might work together to discover these truths. The newspapers were not just a medium for his various causes, whether he was advocating for independence, human rights, celibacy, vegetarianism, religious tolerance, or equality. Gandhi wanted engaged readers. They would not only promote and maintain those causes he treasured, they would also ensure the newspapers were living entities. The Gandhian reader was to read the newspaper, share it in the community, and to write letters to the editor, to which Gandhi would respond. As Gandhian reader, the subscribers to his newspapers had to make a major commitment to the relationship and the enterprise. As Hofmyer notes, Gandhi's readers 'were tasked with almost the same responsibility as the managers of the paper'.[20] Reading *Indian Opinion*, or the newspapers that followed, was an obligation; yet it also was a interviewed the granddaughter, Ela Gandhi, he could not help but come away with the feeling that working for *Indian Opinion* was a higher calling—that indeed it was part deliberate, time-consuming, and inwardly directed activity. It was like studying a discipline, meditating, or praying. Indeed, when the author of this paper and parcel of being a *satyagrahi*. Furthermore, Gandhi saw writing as a way to unite the community through education. For example, during one *satyagraha* campaign, the editor organized a poetry contest on the themes of nonviolence, self-sacrifice, and direct action.[21]

In his book *Hind Swaraj*, we find Gandhi's rhetorical flourishes interwoven into his plain language. For example, in the 'Word of Explanation' at the beginning of the book, Gandhi writes:

> In my opinion it is a book which can be put in the hands of a child. It teaches the gospel of love in place of that of hate. It replaces violence with self-sacrifice. It pits soul force against brute force.[22]

The British rulers of India did not see the appeal of nonviolence and determined that *Hind Swaraj* threatened their power. Thus, government censors banned the book in India. Gandhi stated that the ban 'constitutes further condemnation of the civilisation represented by the British Government'.[23]

Gandhi stated in his autobiography that 'the sole aim of journalism is service', and journalistic writing brought with it a great responsibility to tell the truth. Another responsibility required the editor to interact with his readers. Gandhi also remarked that '*Satyagraha* would probably have been impossible without *Indian Opinion*' because the newspaper gave the reader the most accurate, most intimate account of the campaign.[24]

---

20 Ibid., 133

21 Herman, *Gandhi & Churchill*, 156.

22 Ibid., 15.

23 Mohandas K. Gandhi, *'Hind Swaraj' and Other Writings*, ed. Anthony J. Parel (Cambridge, UK: Cambridge University Press, 2009), 7.

24 Mohandas K. Gandhi, An Autobiography or The Story of My Experiments with Truth (New Delhi, India: Rupa Publications, 2011), 295.

Bibliography

Bhattacharya, Bhanbani. *Gandhi the Writer.* New Delhi, India: National Book Trust, 2002.

Bhattacharyya, S. N. *Mahatma Gandhi the Journalist.* Westport, CT: Greenwood Press, 1965.

Gandhi, Mohandas K. *An Auto biography or the Story of My Experiments with Truth.* New Delhi, India: Rupa Publications, 2011.

Gandhi, Mohandas K. *The Collected Works of Mahatma Gandhi, Volume Two, 1896–1897.* New Delhi, India: Government of India, Publications Division, 1959.

Gandhi, Mohandas K. 'The Grievances of the British Indians in South Africa'. Madras, India: Self-published Pamphlet, August 14, 1896.

Gandhi, Mohandas K. *'Hind Swaraj' and Other Writings.* Parel, Anthony J ed. Cambridge, UK: Cambridge University Press, 2009.

Gandhi, Mohandas K. 'Indian Vegetarians' Series. *The Vegetarian.* London, UK, 1891.

Herman, Arthur Herman. *Gandhi & Churchill: The Epic Rivalry that Destroyed an Empire and Forged Our Age.* London, UK: Arrow Books, 2009.

Hofmeyr, Isabel. *Gandhi's Printing Press: Experiments in Slow Reading.* Cambridge, MA: Harvard University Press, 2013.

Indian Opinion. Durban, South Africa: 1903–1914.

Sanghavi, Nagindas. *The Agony of Arrival, Gandhi: The South Africa Years* (New Delhi, India: Rupa Publications, 2006.

CHAPTER 7

Challenges of Physical and Psychological Spaces Faced by Three Generations of Writers in the 20th and 21st Century in the Sub-Continent, South Asia, and in Pakistan

Syeda Hajirah Junaid

Abstract

The sub-continent was introduced to early forms of texts between 500–300 BC. There is a trivial record of this convergence from text to print over the centuries. However, during all this time spaces had been imperative for writers irrespective to the time or the way of life they were leading. Spaces could and can constitute physical or psychological places that writers want to seek refuge into from the current reality or to exploit the reality that they have been going through. These spaces can be similar if the same profession is followed by successive generations, although it may be completely different from one generation to the next; it could also entirely skip generations. Throughout history, various events have influenced the proletariats as a collective. These events specially have created a prejudice in the spaces created or found by writers. In the 20th Century, the First World War and the Second World War created a bias in the physical, physiological, and psychological spaces of writers. In addition to these Wars, the partition of India and Pakistan in 1947, played an impact in the lives of A.R. Khatoon, my great-grand mother, and her family when her husband went missing in the Second World War and was forced to migrate after the India-Pakistan partition and had to take up writing as her bread and butter for her survival and her children's livelihood. This chapter examines the past and current challenges faced by three generation of writers; A.R. Khatoon, her daughter, and her great granddaughter with respect to their physical and psychological spaces in the 20th and 21st century in the sub-continent and in Pakistan.

Keywords

Zubaida Khatoon – A.R. Khatoon – Pakistani writers – Indo-Pak partition – migration – Spaces – Urdu

•••

 DOI:10.1163/9789004394315_008

1 Spaces

Space is an expansive word and can comprise from a place that one is seated, placed or adjusted to his/her surroundings, vicinities, neighbourhoods to the expanse of his/her city, country or maybe to the universe or his/her galaxy. Spaces can be open areas like a park or the beach however, they can also be enclosed sections like a room or a house. Spaces, be it closed can be public, whereas some other spaces, be it open lands or farms, can be private.[1] The universe and the galaxy will be the same for all writers and so will the desk, chair, bed or room but the psychological spacesfor all mankind will differ. As this will be a collection of past experiences, empirical, and cognitive learning also known as VARK learning.[2]

2 The Need for Spaces

Why does one need space to write? Is there an actual need for such spaces or is it a psychological need? Also, one wonders, do ordinary individuals become writers to grow into existential parallels, instead of living their own life? A lot of the times writers become writers as this is a need for them to become someone or something who they are not and they are, from time to time, transforming in to (that something/someone). Sometimes it is an existence that they form into, unconsciously, however, occasionally their escape becomes so fluid that they become someone who they want to be. At times, it is also an aspiration of the transformation to become a character, live in that particular household or virtually transport from a world of despair that they want to escape, into their subconscious from the conscious.

3 Shift in Refuge: from Reading Books to Writing Books

Writing becomes imperative when the already present text of books, articles, newspapers and other materials do not satisfy a writer. There seems to be a

---

1 'Definition of Space/s in English', *Oxford Dictionaries*, Accessed 1 June 2016, http://www.oxforddictionaries.com/definition/english/space.

2 Susannah Shatuck, 'The Four Different Types of Learners and What They Mean to Your Presentations [Infographic]', *Prezi Blog*, Accessed on 28 May 2016, https://blog.prezi.com/the-four-different-types-of-learners-and-what-they-mean-to-your-presentations-infographic/.

need to write something that doesn't already exist but can only exist if the writer puts pen to paper. It always starts with a writer's desire to read a specific kind of story/ characters that are ever-present in his mind. This need to satisfy that desire and make his/her stories available for others to see and read, is when the writer starts on his journey to write.[3] There are times when stories are said to consume the individuals and it becomes a duty, an obligation to put pen to paper but it is not a simple task as the story, the characters, the journey start to become so involved with the individual that it seems to penetrate his/ her conscious mind.[4] Psychological presence of the book, story or fable becomes larger than life which needs to be written in such a way that it forms characters, beings, and individuals; that needs to be presented to other individuals.

4 The Need to Write

Mundane tasks of everyday become necessary to escape especially for writers. Yet this is only the psychological aspect of it, there are physical and monetary needs that have to be fulfilled in such a way that writing becomes a task, a job, a way to earn a livelihood and there is no need or time to be absorbed in the characters.

This chapter examines the need for writers to write; in particular, A.R. Khatoon (my great-grandmother), her daughter Zubaida Khatoon (my grandmother) and the spaces for writing in the 21st century, as I take on the challenge to write. It is imperative to understand the reasons why writers write. There is a theory that explains the physiological and psychological needs that every human possesses. It is called Maslow's Theory of Needs[5] and it helps us understand

3 Maria Popova, 'The Six Motives of Creativity: May Gaitskill on Why Writers Write', *Brain Pickings,* Accessed on 29 May 2016.

4 Saadat Hassan Manto, 'Why I Write Short Fiction Stories', *Ravi Magazine,* Accessed on 26 May 2016, http://www.ravimagazine.com/why-i-write-short-fiction-stories-main-afsana-kyonkar-likhta-hoon-manto/.

5 A.R, Khatoon, Full Name: Amt-ur-Rehman Khatoon. 1900–1964 Parents passed away in early ages then lived with Grandparents in Makkah. Her father was Head of Islamic Education division in India, mother was a housewife. Lived in Makkah till her adolescence. Got engaged but due to some widespread illness, Fiancé, and grandparents died in Makkah. Travelled back to India and got married to Fiancé's brother. Published 8 books in Urdu including Shama, Tasweer, Afshan, Chashma, Hala,Romana, Faakiha and Kahanian. A.R. Khatoon's novels during 1930s-1950s made it possible for females in India and Pakistan to have ambitions and dreams to pick an occupation and have a role in the society. Her books made it acceptable for woman to take part in the male dominated society of the Indian subcontinent and yet have a good standing in her family. Beauty and money were the heredity of women of the east however,

the need for writers to write. In this theory, Maslow described the basic needs every human being has and relates that to their specified Social Economic Strata. He claims that when physiological needs are met, like food and shelter, human beings not only tend to climb the ladder of psychological needs but also have a desire to address them. The lowest need is the Physiological need, it includes food, water, shelter; these are the hygiene or animalistic needs that require priority. When the lowest need for survival is met, then human beings tend to look for other requirements. This Physiological Need served as an obligation for A.R. Khatoon as she became a Bread Earner in 1942 (when her husband went missing in the Second World War and then killed in 1945) by writing (publishing in Urdu) to support her family. Her need became a Safety Need when she started receiving pension for her deceased husband. Before the war in the 1930s, when she wasn't the sole earner of her family, she wrote to address her need of Self-Actualization (this is the highest psychological need) when she created characters and stories to fulfil her desire to fictionalize a parallel universe.

Her daughter Zubaida Khatoon[6] was not solely dependent on her own writing (publishing in Urdu) financially as she would receive income from her husband. However, as her children became older and their expenses increased, her need to write became a Safety Need in her later years. She also claimed to write for fame and to improve her self-esteem by writing. To understand other needs, a comparison on why other writers wrote has been done. J.R.R. Tolkien wrote 'The Hobbit' and 'The Lord of the Rings' to narrate an attractive story to his children which categorizes him in the Love/ Belonging Need. However, it is said that his need to create stories mixed with a need to experiment with German and Norfolk elements of culture, myth and traditions along with some Gothic components, made it a need of Self-Actualization, where his need could not be fulfilled with the existing literature.[7] Jane Austen was said to write to

---

A.R. Khatoon tried to inculcate the importance of education in the lives of woman and how this could benefit from the medieval dogma traditions that was being followed by the woman in those times. Her psychological spaces demanded that her characters strongly believe that getting education in English should become a must for all the individuals of that time and women should be educated at least till a Bachelor's degree.

6 Zubaida Khatoon, 1926–1980; Parents: Housewife and Tax Collector MIA in Burma in 1942 and was declared dead in 1945. Eldest of 5 siblings. Author of 8 books, 2 of which were converted to Drama Serials on TV. Influenced by Partition of Indo-Pak 1947, the Second World War, 1965 and 1971 War fought with India. Published books in Urdu include: Huma, Aroosa, Kiran, Faakiha, Nihaala, Aini, Nadra and Tarunnum. Comparison given in the appendix below.

7 Humphrey Carpenter, *Biography of J.R.R. Tolkien* (London: Unwin Paperbacks 1978).

become independent.[8] A.R. Khatoon was coined as the 'Jane of Austen of the sub-continent' which shows how similar both these writers were.[9] Both wrote about family lives and how the clever characters took advantage of situations, related to money, fame and marriage.[10] Both Jane Austen and A.R. Khatoon had strong but complex female characters in their books.

5 Cognitive Models for the Process of Writing

Cognitive models define writing in terms of problem solving.[11] Writers often follow a strategy known as knowledge transforming when they *problematize* any writing task. Expert writers tend to develop elaborate goals or artificial goals and tend to problem solve the content towards that particular goal. Inexperienced writers or novice writers tend to have a simpler approach of storytelling where every step of the content follows the other.[12]

A.R. Khatoon had an eye for problem solving and was a good chess player. She was also able to decipher many situations which would seem unsolvable to the human eye. She solved many myths that seemed super natural to the common man but her logical and rational thinking brought order to the chaos. This is the reason why her books are complex and have characters which are interwoven with historical facts.[13]

6 Challenges

The challenges that writers face in their daily lives to write seem plentiful considering the writing process and the need for the writing to be consistent with mind to paper, and the flow of the story. The images in the mind would seem banal and would exist only in the mind of the writer if not put to words in such

8 Richard Bentley, 'Biography: Life (1775–1817) and Family of Jane Austin', *Pemberley*, Accessed on 20 May 2016, http://www.pemberley.com/janeinfo/janelife.html.

9 Khatoon, *Adeeba*, 172–255.

10 Khatoon, *Adeeba*, 172–255.

11 Paul Deane, Nora Odendahl, Thomas Quinlan, Mary Fowles, Cyndi Welsh, JenniferBivens-Tatum, 'Cognitive Models of Writing: Writing Proficiency as a Complex Integrated Skill', *ETS*, Accessed on 18 May 2016, https://www.ets.org/Media/Research/pdf/RR-08-55.pdf.

12 Dean Paul, Odendahl, Quinlan, Fowles, Welsh, Bivens-Tatum, 'Cognitive Models of Writing'.

13 Zubaida Khatoon, *Adeeba: A Biography of A.R. Khatoon* (Lahore: Ashraf Press, 1967).

a way that the mind of the reader is envisioning it with vivid clarity or understanding the same perceptions.

The mundane distractions of everyday life bring frustration to writers and their writing process. That was the case of Zubaida Khatoon who seemed perplexed by daily tasks, inadequate temperatures of the house in the 1960s and 1970s, prolong diseases, lack of availability of paper (especially during the war days) and hordes of visiting relatives. Her process of writing became very erratic due to these issues. Zubaida Khatoon would write many pages after purchasing paper but would consider her work irrelevant due to lack of discipline in producing the flow. Yet the lack of finances and the escape from the prosaic life brought a sense of motivation that compelled her to write year after year.

7 Impact of Spaces

The consequence of having conscious or subconscious physical and psychological spaces provides refuge for writers. This sanctuary makes it possible for writers to be deliberately motivated to live in the present while creating a parallel life in their mind. In some cases, like in the case of A.R. Khatoon, it seems to resonate her introverted personality and her unconscious way of living in an ivory tower of complacency especially when she started her first novel 'Shama'.[14]

The following were the different types of impacts of spaces which explain how they affects generations of writers.

7.1 *Impact of Spaces on Successive Generations*

The physical space in which these writers of Indo-Pak wrote, comprised of the solitude of their homes. A.R. Khatoon wrote in her room during the afternoons. Zubaida Khatoon is said to take refuge in writing at a dinner table, in the living rooms and sometimes on the bed depending on how many guests were staying at her place, which varied from one to twenty people daily. Both A.R. Khatoon and Zubaida Khatoon travelled a lot due to father's/husband's posting and spent their summer vacations at their parents' place. This gave them an in-depth insight into the life of various towns, villages and cities which made it easy to communicate in their books.

The familiarity of having a parent as a writer came as an advantage to Zubaida Khatoon. She grew up in an environment of an introvert mother, who had a logical mind and could create anecdotes out of thin air. Zubaida Khatoon

14 A.R. Khatoon, *Shama* (Karachi: Welcome Book Point, 1994).

would often invade A.R. Khatoon's space and would ask her relentless questions about how her mother could write and what were the key elements in her writing. Despite understanding this, not only did it bring greater responsibility and pressure to Zubaida Khatoon to carry the legacy of her mother but also forced her to step into her mother's shoes when she had to complete her mother's unfinished book *Faakiha*. Zubaida Khatoon was very disturbed in 1966 (a year after her mother's demise) when she had to fulfil her mother's dying wish of completing this last book. She wanted to understand her mother's psychological space by re-reading her previous books but was weary of the burden, as not only was she in grief but was also forced to re-live her mother's existence by completing this last book. She finally completed it by the insistence of her agent and it was published in 1966.[15]

7.2 *Impact of Spaces in the 20th Century*

The 20th century was dominated by wars in many countries of Europe and Asia, and the disaster as the aftermath of the war. Despite the war being fought by soldiers and officials in combat that took place in water, air and on the ground, many wars were fought at home by the bystanders and onlookers. This reverberated with the writers of the 20th century and created a deep impact on the characters created by them. The inspiration of Sam Wise Gamgee in 'Lord of the Rings' is said to have come about by the helpers/ runners of the First World War.[16]

There are repeated mentions of the Indo-Pak partition of 1947 in the books of Zubaida Khatoon and A.R. Khatoon. These bear a similarity to their own story of migration from Delhi to Lahore as mentioned in a letter by A.R. Khatoon.[17] *Faakiha*, the book completed by Zubaida Khatoon, narrates the story of a young girl and her family. This is a story of how she has to migrate with her parents despite lack of security in the country. It also tells us how her family preferred to migrate from India to Pakistan via an airplane instead of a train as trains had become slaughter houses for the radicals.

Same was the case when individuals from Pakistan fled to India. Both A.R. Khatoon and Zubaida Khatoon have several protagonists serving during the

15 Zubaida Khatoon, *Unpublished Diaries*, 1966, 1970,1971, 1972, 1973, 1975, 1976, 1977, 1978, 1979, 1980.

16 Carpenter, *Biography of J.R.R. Tolkien.*

17 She wrote to the editor of *'Ismat'* magazine to explain the reasons for her not writing. She elaborated on the dilemma that she was drowning in financial troubles, lack of justice received in allotment of a house and the problems of Partition of Indo-Pak. A.R. Khatoon's process of writing included her refuge in the quiet summer afternoons when her children were forced to sleep and she wasn't accountable for the responsibilities.

war like Zubaida Khatoon's father. There are mentions of the refugee camps in Pakistan in which migrants came in and stayed after the partition of 1947 which represents the lack of brick and mortar properties for even the elite and the struggles that they went though.[18]

Other than wars, there are mentions of great events like the earthquake that devastated Quetta, Pakistan in 1930. The mention of this earthquake in A.R. Khatoon's historical fiction based book 'Shama', served as a big plot twist that left a young girl orphaned and in the hands of her cunning relatives.

Also, death seems to be a common thread amongst the writers of the 20th century. Since the presence of wars and the ease with which death prevailed in this century, it becomes acceptable for writers to involve death in the books and make it impressionable on the characters. Mostly death was supposed to be a learning curve for the virtuous protagonist's family, after which he or she would take on the challenges of the world with hardihood.

However, like other individuals who have been through wars, writers also tend to loathe wars with passion. Zubaida Khatoon's father went missing in the Second World War and remained missing for three years, to be declared dead in 1945. This not only brought physical and mental trauma to his family but it remained with the children till their end. Being anxious, restless, disturbed and apprehensive seemed to resonate throughout Zubaida Khatoon's diaries.[19] This followed with an immediate feeling of losing family members every time one left the house. However, writing also brought an escape to her troubled mind. Similarly, J.R.R. Tolkien mentions:

One has to indeed personally come under the shadow of war to feel fully its oppression; but as the years go by it seems now often forgotten that to be caught in youth by 1914 was no less hideous an experience than to be involved in 1939 and the following years. By 1918 all but one of my close friends were dead.[21]

7.3 *Spaces in the 21st Century*

Physiological spaces have come a long way since the 20th century to the 21st century. Long gone are the 2 feet pencils that writers like A.R. Khatoon used to write 600 page novels. The current writer is equipped with modern technology

18 A.R. Khatoon, *Chashma* (Lahore: Welcome Book Point, 1958). There are several mentions of vacant houses by previous owners in her books, who fled to save their lives leaving all their furniture, clothes, utensil, everything intact even leaving food on the stoves. It represented the helplessness of these physical spaces and how man is made flexible to adjust to unfamiliar areas. All these were taken from their real-life experiences.

19 Zubaida Khatoon, *Unpublished Diaries*, 1966, 1970–1980.

of computers and can use it anywhere at any time. During long hauls of light outages or without having to worry about the shortage of paper, these technologies can be used with convenience. However, the psychological spaces, perhaps remain the same. Anger, frustration and emotions of despair bring out the best in writing, especially by novice writers like myself.

Everyday life is improving as more technology and rationale is being taught through education. However, the same power struggle over money and dominance from the lowest levels of the society to the highest levels seems to take over the world resulting in corruption and sometimes even wars. This brings us all a full circle to the same point our ancestors were standing at fifty to hundred years ago, hence creating the same psychological spaces that they went through.

8 Conclusion

Centuries seem irrelevant to teach mankind lessons that they keep on repeating. The frustrations of writers seem to be alike even after decades. The mode of communication may seem faster but the problems are the same. Females have more empowerment, as was the hope of A.R. Khatoon, but humanity as a whole is still struggling with similar issues. Paper, pencils, pens and the corners of a room are shrunk to laptops, tablets and phones, however, the blank spaces in the minds remain the same.

The impact of wars, migration and displacement serves as a lifelong brutal mark on the minds of individuals, especially writers, and are a constant reminder, generation after generation and make them sensitive just like before. Not only does it stay for that generation, but it also transfers to the next generation through various thought processes. This impact is reflected in the way people, especially writers, live their lives, think about the future and overcome the horrors of the past. Writers mould this experience in their stories and take it forward when creating new experiences for the readers. No matter what time, century or region the writer lives in, they consciously or subconsciously live in parallel dimensions of their writing.

Bibliography

Bentley, Richard. 'Memoir of Jane Austin'. *Pemberley*. Accessed on 20 May 2016. http://www.pemberley.com/janeinfo/janebblg.html#memoir.

Burrows, John. 'Computation into Criticism: A Study of Jane Austen's Novels, and an Experiment in Method'. *Library Catalogue, Oxford University Press 1987*. Accessed on 20 May 2016. http://lib.ugent.be/catalog/rug01:000132584.

Behrens, Katie. 'J.R.R. Tolkien's Epic Quest: Writing *The Lord of the Rings*'. *Books Tell You Why* [*Blog*]. Accessed 10 May 2016. http://blog.bookstellyouwhy.com/jrr-tolkiens-epic-quest-writing-the-lord-of-the-rings.

'Definition of Space/s in English'. *Oxford Dictionaries*. Accessed 1 June 2016. http://www.oxforddictionaries.com/definition/english/space.

Humphrey, Carpenter. *Biography of J.R.R. Tolkien*. London: Unwin Paperbacks 1978.

Khatoon, A.R. *Chashma*. Lahore: Welcome Book Point, 1958.

Khatoon, Zubaida. *Adeeba: A Biography of A.R. Khatoon*. Lahore: Ashraf Press, 1967.

Khatoon, Zubaida. *Unpublished Diaries*. 1966, 1970, 1971, 1972, 1973, 1975, 1976, 1977, 1978, 1979, 1980.

Manto, Saadat Hassan. 'Why I Write Short Fiction Stories'. *Ravi Magazine*. Accessed on 26 May 2016. http://www.ravimagazine.com/why-i-write-short-fiction-stories-main-afsana-kyonkar-likhta-hoon-manto/.

McLeod, S.A. 'Maslow's Hierarchy of Needs'. *Simply Psychology*. Accessed on 20 May 2016. www.simplypsychology.org/maslow.html.

Melodie, Neal. 'What Did J.R.R. Tolkien Like to Read?'. *Quora*. Accessed 10 May 2016. https://www.quora.com/What-did-JRR-Tolkien-like-to-read.

Moody, Ellen. 'Jane Austen's Heroes, and Austen's *Sir William Mountague*'. *Pemberley. Accessed on 10 May* 2016. http://www.pemberley.com/janeinfo/emaushro.html#partC.

Paul Deane, Nora Odendahl, Thomas Quinlan, Mary Fowles, Cyndi Welsh, Jennifer Bivens-Tatum 'Cognitive Models of Writing: Writing Proficiency as a Complex Integrated Skill', *ETS*. Accessed on 18 May 2016. https://www.ets.org/Media/Research/pdf/RR-08-55.pdf.

Popova, Maria. 'The Six Motives of Creativity: May Gaitskill on Why Writers Write Art of Integrating the Ego and the Impulse for Empathy in a Dynamic Call and Response'. *Brain Pickings*. Accessed on 29 May 2016.https://www.brainpickings.org/2014/02/28/mary-gaitskill-writing/.

Dr. Robina Shehnaz, Dr. Naeem Mazhar, Dr. Zafar Ahmed, Dr. Nazia Malik. 'Daryaft'. NUML Educational Paper. *NUML Documents*. January. 2016. Accessed on 20 May 2016. http://www.numl.edu.pk/docs/Daryaft%2015%20final.pdf.

Shatuck, Susannah. 'The Four Different Types of Learners and What They Mean to Your Presentations [Infographic]'. *Prezi Blog*. Accessed on 28 May 2016. https://blog.prezi.com/the-four-different-types-of-learners-and-what-they-mean-to-your-presentations-infographic/.

Tolkien, J.R.R. *Lord of the Rings*. London: HarperCollins, 2004.

Appendix

Jane Austen	J.R.R. Tolkien	Zubaida Khatoon	A.R. Khatoon
Birth: 1775, 7th child (2nd daughter) *Parents*: Married and alive Started writing 1795. Attended formal schooling only for 1 year	*Birth*: 1892, South Africa *Parents*: Father died when he 3 was and mother when he was 12. A family friend Father Francis Xavier was his guardian	*Birth*: 1926 *Parents*: Housewife and Tax Collector MIA in Burma in 1942 and was declared dead in 1945. She had a Scottish Governess who returned to Scotland due to WWII in 1938–39. Attended school till class 13	*Birth*: 1900 Parents passed away in early ages then lived with Grand parents in Makkah. Father was Head of Islamic Educated division, mother housewife. Didn't attend formal schooling
Places lived: Hampshire, Reading, Southampton, Bath, Chawton	*Places lived*: Birmingham, Sarehole, Worcestershire, Switzerland, and South Africa	*Places lived*: Gonda, Behraich, Bara Banki, Delhi, Aligarh, Faizabad, Shah Jahanpur, Simla, Karachi, Lahore, Taxilla, Rawalpindi, Hassanabdal, Multan and Bahawalpur	*Places lived*: Makkah, Bhopal, Gonda, Behraich, Bara Banki, Delhi, Aligarh, Faizabad, Shah Jahanpur, Shimla, Karachi and Lahore,
Influenced by: Her family life, her house, family events	*Influenced by*: World War I and II and the experience that brought out for Sam Wise. Atomic Bomb and The Ring's similarity	*Influenced by*: Partition of 1947, the II World War, 1965 war, Fall of Dhaka	*Influenced by*: the earthquake in Quetta, Influence of the British Rule
Reasons for writing: Started to write for Leisure then from 1793 she started to write professionally	*Reasons for writing*: LOTR was written as the writer seems to want to narrate that attractive tale also wanted to experiment between European cultures and languages	*Reasons for writing*: Started off as a passion but later converted to a need to earn money. 1. Fame 2. Money 3. Escape	*Reasons for writing*: Started off as a passion in 1920s. Short stories and Shama was a passion but the rest of them were to earn money as she became the bread earner

Jane Austen	J.R.R. Tolkien	Zubaida Khatoon	A.R. Khatoon
Problems: When her father retired and moved the family to Bath in 1800, Jane Austen was disturbed and maybe in depression in 1805 after the passing of her father	*Problems*: Trauma of losing parents really early. Troubled by war and lack of resources especially lack of paper	*Problems*: Trauma of father, Migration, Lack of stability in the country, lack of resources especially lack of paper. Unavailability of a proper house	*Problems*: Husband MIA. Migration. Being a female bread earner in a new country. Allotment of a disputed house
Friends: Cassandra (her sister)	*Friends*: C.S. Lewis	*Friends/ Acquaintances*: Bilqees Zafar, Zakia Akbar	*Acquaintances*: Altaf Fatima, Bilqees Zafar, Haqeem Saeed,
Influence: Sir Charles Grandison, William Wordsworth, Samuel Coleridge and Lord Byron. Plays: Richard Sheridan and David Garricks	*Influence*: Beowolf, Shakespeare, Charles Dickens, William Morris, Andrew Lang, George Mac Donalds, Owen Barfield, H. Rider Haggard, Samuel Rutherford, Richard Wagner, S.R.Crockett	*Influence*: Fiaz Ali, his novels Shamim and Anwar, Ibn-e-Safi Novels and series. Monthly magazines Woman's Own , Lail-o-Nihaal, and Urdu Digest	*Influence*: Fiaz Ali, his novels Shamim and Anwar. Also influenced by modernism
Published work: Pride and Prejudice, Northanger Abbey, Persuasion, Sense and Sensibility, Emma, Mansfield Park, Lady Susan, Unfinished: The Watsons and Sandition, Juvenilia Volume I, II and III	*Published work*: Beowolf, Letters, The Hobbit, The Lord of the Rings, The Silmarillion, The Unfinished Tales and The History of Middle Earth, Mr. Bliss, Legend of Sigurd and Gudurn, The Story of Kullervo	*Published work*: Books including Aroosa, Kiran, Faakiha, Romana, Nihaala, Tasweer, Aini, Nadra, Tarunnum. Voice for her books on Radio Pakistan	*Published work*: Short stories in various Magazines including Ismat etc. Books including Shama, Afshan, Haala, Kahanian, Chasma

CHAPTER 8

Writing as the Philosopher: Murdoch and the Theory

Barış Mete

Abstract

The Anglo-Irish novelist Dame Jean Iris Murdoch (1919–1999) published twenty six novels starting with *Under the Net* in 1954 and ending with *Jackson's Dilemma* in 1995 just four years before her death. Three of these novels have been awarded literary prizes including the Booker Prize for one of her most acclaimed works, her 1978 novel *The Sea, the Sea*. Publishing a novel almost in every year in the second half of the twentieth century, Murdoch was so prolific as a novelist that her theoretical writings as a philosopher have not been given the same critical attention. Although Murdoch began her career as a novelist, she had studied philosophy between 1947 and 1948 at Cambridge. After her study of philosophy as a postgraduate, Murdoch had actually taken up a fellowship and thought philosophy at Oxford until 1963. It could now be argued that Murdoch was essentially a Platonist philosopher who commenced writing fiction in an attempt to put into practice her own theories of art and literature. It is true that Murdoch made in her writings a broad outline of the theories of Plato, yet she at the same time particularly reinterpreted the reasons why Plato had forcefully accentuated in his philosophical discussions an adverse criticism about literature as a form of representative arts. This chapter will accordingly illustrate through examples how a novelist philosopher rephrases and rewrites the part of classical literary theory both in her fiction and in her philosophical writings.

Keywords

Writing – rewriting – classical philosophy – theory – art – fiction – fictional theory

•••

 | DOI:10.1163/9789004394315_009

1 What Is Art?

The canonical Western theory of art has been based on the principal that art is an imitation, and thus all forms of art are regarded as representations. This basic assumption is a reflection of the theoretical illustrations of not only Plato but also Aristotle who obviously agreed on the decisive conclusion that art was a mimetic profession. Although Plato, whose discourse on art can thoroughly be summed up in his assertion that 'art is always bad for us in so far as it is mimetic or imitative,'[1] associated imitation with untruth and deceptiveness, his most famous student, Aristotle, who as Murdoch states 'was not frightened of art,' and who as she further claims 'held that presentation of life in art could have an ennobling effect,'[2] believed that imitation was part of human nature and men learned life through imitation. According to the Aristotelian theory of art, as it has been, as a term, more widely acknowledged to refer to the mimetic theory of art today, art has a mimetic nature and it copies the source material – life, reality, or nature. As an artist, but more particularly as a novelist, Murdoch, who 'has still the distinction of being designated as a philosopher-novelist, as if the writing of fiction were for her a holiday from the more taxing and responsible profession of philosophy,'[3] and whose thought is said to have 'renewed interest in Platonic ethics and virtue theory, a reopening of the debate over the relation between philosophy and literature,'[4] believes in accordance with the mimetic theory of art that 'art imitates nature – art is the imitation of nature.'[5] Since Murdoch was a novelist who was a Platonist philosopher as well, her theories of literature have substantially characterised the reworking of the particular examples depicted especially by Plato in his own discussions. For example, as a Platonist, 'What is distinctive about Murdoch's theory of the novel is that it is also a theory of ethics.'[6]

As it is well known, in the tenth chapter of his *Republic*, Plato talks about the harmful effects of representative arts on the members of a healthy political system where it is essential that 'each entity becomes aware of the other and realizes that one's existence severely depends on the wellbeing and the

1 Iris Murdoch, 'The Fire and the Sun: Why Plato Banished the Artists,' *Existentialists and Mystics: Writings on Philosophy and Literature*, (New York: Penguin Books, 1999), 390.

2 Iris Murdoch, 'Art is the Imitation of Nature,' *Existentialists and Mystics: Writings on Philosophy and Literature*, (New York: Penguin Books, 1999), 243.

3 Iris Murdoch, 'Art is the Imitation of Nature,' 247.

4 John Sturrock, 'Reading Iris Murdoch,' *Salmagundi*, 80 (1988): 144.

5 Maria Antonaccio, 'Imagining the Good: Iris Murdoch's Godless Theology,' *The Annual of the Society of Christian Ethics*, 16 (1996): 223.

6 Murdoch, 'Art is the Imitation of Nature,' 243.

existence of the other'[7] for he identifies all forms of artistic representation especially with deception, and thus falsehood. Plato inaugurates his Socratic dialogue in the specified chapter with the poet, yet he concentrates more on the painter for he explains his central idea, the notion of untruth, more illustratively through the image of a painting on which, Plato absolutely believed, what appeared was no more than an illusion. According to Plato, the painter – and on the condition that this will only be metaphorical – will

> soon create the sun and the heavenly bodies, soon create the earth, soon create yourself, other living creature, furniture, plants and all the things we've just been talking about ... But not ... as they truly are ... Something *like* the real thing, but not itself the real thing.[8]

Similar to what Plato did, Murdoch refers to the same discussion about the painter and claims the idea that 'painting often serves as a sort of explanatory metaphor for the other arts.'[9] Furthermore, Murdoch believes that Plato's focus on the art of painting is because of the fact that the painting is a very clear and typical example. In addition to this, she explains Plato's use of the image of a painting in the idea that human mind is conditioned to reproduce the original model in another medium. Therefore, Murdoch argues that the painter or the painting becomes the point of reference for every viewer for

> the image of the painter who sits there at his easel and looks at the landscape is a profound idea, which we carry with us as we think of different forms of art.[10]

Plato, according to Murdoch's interpretation, believed that the painter as an imitator transformed what he looked at – for example the landscape – into an illusion whose character was no longer the same as that of reality itself – the landscape. Although the painting will inevitably be a reproduction that is yet similar to reality, what the painter produces will certainly be a product that is completely different in itself.

---

7 Bran Nicol, 'Iris Murdoch's Aesthetics of Masochism,' *Journal of Modern Literature*, 29.2 (2006): 148.

8 Huseyin, Altindis, 'The World Will Go On Living: Resistence to Eurocentric Epistemology in Linda Hogan's *Power*,' *Interactions: Ege Journal of British and American Studies*, 26.1–2 (2017): 22.

9 Plato, *The Republic: Cambridge Texts in the History of Political Thought*, G.R.F. Ferrari (ed.), Tom Griffith (trans.), (Cambridge: Cambridge UP, 2000), 315.

10 Ibid.

Plato not only denounced representative literature as imitation but he also believed that the poet as the imitator was a manipulator for 'the imitator will have neither knowledge nor correct opinion about the goodness or the badness of the things he imitates.'[11] The poet, very much similar to the painter, creates only illusions that appeals to 'the ignorant majority'[12] for 'the imitator has no knowledge worth mentioning of the things he imitates.'[13] While imitation as a notion was categorically unwelcome for Plato since he believed that 'poets do not know what they are doing and they cannot explain their work,'[14] Murdoch specially makes a historical reference to the art of sculpture in classical Greek culture. According to her reference, the existence of the illustrious examples of this art form in ancient Greece practically validates the deep-rooted situation of imitation in the society. Murdoch argues that

> The Greeks spent a hundred years perhaps of the best time in the history of sculpture, as we know it, in trying to learn how to imitate the human form. This is an interesting case of a pure effort of imitation.[15]

Aristotle's definition of literature as imitation – and thus the point where he agrees with Plato – includes epic poetry, tragedy, comedy, dithyrambic poetry as literary types as well as music as flute playing and lyre playing.[16] In other words, Aristotle observes that music as an art is a product of imitation. The difference between the literary types and music, for Aristotle, is that not only do they imitate different things but also they imitate quite differently.[17] In order to interpret the meaning and the significance of the notion of imitation in the Aristotelian definition, Murdoch, whose principal question is what exactly imitation is and what counts as imitation,[18] first attempts to illustrate the relationship between music and imitation. At this point, however, Murdoch thinks that she is unable to discover any sign of true connection between music and imitation. It would otherwise be so absurd that she says, 'music doesn't seem to imitate nature except in some rather trivial sense.'[19]

11 Ibid.
12 Plato, *The Republic*, 322.
13 Ibid.
14 Ibid.
15 Murdoch, 'Art is the Imitation of Nature,' 246.
16 Murdoch, 'Art is the Imitation of Nature,' 243.
17 Aristotle, *Poetics*, Joe Sachs (trans.), (Newburyport: Focus Publishing, 2006), 19.
18 Ibid.
19 Murdoch, 'Art is the Imitation of Nature,' 243.
Murdoch, 'Art is the Imitation of Nature,' 244.

Here Murdoch makes a comparison between two art forms, music and painting, assuming that her illustration will help uncover the reason for her conclusion that music is hardly an imitation. Compared to painting, Murdoch maintains, music does not create the same effect of verisimilitude on the listener. As a result of this, she refrains from elaborating on the Aristotelian reference to music as imitation. As Murdoch continues concentrating more on painting, she introduces another theme, realism in art, related to the main topic of her discussion, art as imitation. A map and a countryside, according to Murdoch, are not similar even if 'we move a little bit away from art but remaining within a convention which is certainly related to art.'[20] A map does not look like a countryside due to the obvious lack of realism in the former whereas any artistic representation of the latter in Aristotelian logic requires a realistic depiction.

2 Realism and Imitation

As it has always been the case in the theory of representative arts and literature, realism itself – which would perhaps be considered as a solution to comprehending imitation – is indeed a puzzling term for Murdoch as well. It is so puzzling that even the most realistic art is conditional on the fact that 'If one takes fairly ordinary and familiar cases it is clear that "x" looks like "y" is subject to all sorts of conventions and historical changes.'[21] In other words, the idea of resemblance between two objects is defined by some broader and more encompassing situations rather than a simple image of likeness. As a result, Murdoch argues that realism as a term is insufficient to describe imitation even if the examples are to be chosen from painting. This is because '"The impressionist paints what he sees, the Dutchman paints what he thinks."'[22]

Murdoch asks who the most realistic novelist is and believes it is even in literature that realism is impossible to define. This is a consequence of the fact that all literature – necessarily including both realistic and experimental examples – is imitation. According to her, Jane Austen, Gustave Flaubert and Charles Dickens were realistic writers for these writers imitated the physical and the social reality they were part of. Murdoch adds the idea that Virginia

20 Ibid.

21 Ibid.

22 Ibid.

Woolf and Alain Robbe-Grillet as experimental writers were imitative writers as well for they imitated the same source as Austen, Flaubert and Dickens did.[23]

If art is fundamentally a product of imitation, and if what imitation means here is the imitation of nature (i.e. reality), then why are experimental art forms (e.g. experimental literature) not considered realistic art? This is because criticism has, as Murdoch underlines, left out an important issue, the issue of 'aesthetic merit.'[24] Imitation for Murdoch thus is an artistic endeavour unquestionably characterised by its aesthetic value and appeal. Furthermore, it is the aesthetic significance of art that art has frequently been associated with truth. In terms of the relationship between truth and art, Murdoch reinterprets the Platonist notion of the painter painting the bed. She says, 'To revert to the case of the bed, the painter can reveal far more than the "one viewpoint" of the ordinary observer.' And she adds that 'The painter and the writer are not just copyists or illusionists, but through some deeper vision of their subject matter may become privileged truth-tellers.'[25] Murdoch makes another point 'which, in fact, has already been said by Keats, that Art is Truth, art is some kind of truth, art is true.'[26] This might be the reason why, for example, Murdoch's artist protagonist in her 1973 novel *The Black Prince* combines art – and thus his position as an artist – with truth. He says, 'Good art speaks truth, indeed *is* truth, perhaps the only truth.'[27]

The combination of art and truth, however, might be originated from the suggestion that '"We learn from art, we learn to look at the world, art is a kind of statement, it has a strong cognitive aspect."'[28] The idea that art teaches, or that art is a medium for education, as Murdoch cites here, comes from the Aristotelian definition of poetry – poetry as representative art – where imitation as a natural ability of man is designated as a way of learning: 'imitating is co-natural with human beings from childhood,' Aristotle says, and he adds that 'in this they differ from the other animals because they are the most imitative and produce their first acts of understanding by means of imitation.'[29] If art necessarily teaches the truth, if man learns the truth from art, then the conclusion of this idea would be the assumption that good art is true whereas bad art is false.[30]

23 Ibid.
24 Ibid.
25 Murdoch, 'The Fire and the Sun,' 392.
26 Murdoch, 'Art is the Imitation of Nature,' 244.
27 Iris Murdoch, *The Black Prince*, (New York: Penguin Classics, 2003), 3.
28 Murdoch, 'Art is the Imitation of Nature,' 245.
29 Aristotle, *Poetics*, 22.
30 Murdoch, 'Art is the Imitation of Nature,' 245.

3 Criticism as Hostility

It is obvious that Plato as a philosopher and as a literary theorist was hostile to representative arts. For Murdoch, however, such hostility to art, but particularly to literature, is not limited to Plato; but it has extended to the modern times and is practiced today in the form of modern literary criticism. Sontag, for example, interestingly writes about the issue that 'The modern style of interpretation excavates, and as it excavates, it destroys.'[31] Murdoch correspondingly says, 'Some criticism of literature is purely formal, but very much more of it is … moral, and … the critic may accuse the writer of some kind of lying or misrepresentation.'[32] In other words, literary criticism might possibly recognise falsehood in literature. As it has been illustrated, the discussion has so far concentrated on imitation and the classical interpretations of imitation as the essence of art; and as it has further been pointed out, Plato blamed the poet for being an imitator. Murdoch hence outlines Plato's notion of imitation – or mimesis as it is the original word – as the source of his hostility to representative arts in three main reasons. According to Murdoch's interpretation,

I. Plato 'was afraid … of the power of art as persuasion or propaganda, art as irrational emotion, art as attractive lies … art as the telling of inconvenient or subversive truth.'[33] Although such a remark as art as propaganda (Murdoch sees Plato's approach here as a 'utilitarian theory of art.'[34]) quickly reminds the listener of the teachings of the part of Marxist criticism which has forcefully suggested that literature should serve the interests of the party,[35] Plato attempts, not as a literary but as a political theorist here, to protect the political structure and the order of the Greek city state. For Murdoch, however, the analogy between the Platonist and the Marxist theories are significant enough to assert that 'Some Marxists think that literature should provide immediate pamphlets or improving stories to serve the present state of the revolution'[36] and she believes that this is similar to what Plato did. As art is an illusion for Plato, it thus has a deceptive nature that could impair the minds of the citizens of a healthy state.

II. Plato 'regarded art as a prime distraction from the study of nature. Art would inhibit reflection and liberating thought. Goodness and knowledge, for

31 Susan Sontag, 'Against Interpretation,' *Against Interpretation and Other Essays*, (New York: Penguin Classics, 2009), 4.

32 Murdoch, 'Art is the Imitation of Nature,' 245.

33 Ibid.

34 Murdoch, 'Art is the Imitation of Nature,' 248.

35 Terry Eagleton, *Marxism and Literary Criticism*, (London: Taylor&Francis, 2006), 19.

36 Murdoch, 'Art is the Imitation of Nature,' 248.

Plato, both lay in the same direction.'[37] The greatest artist for Plato is God, and thus all other artists are not creators but mere imitators who imitate either God (the carpenter) or the carpenter (the poet or the painter). As a result, the representative artist has no connection with truth (Ideas or Forms). Besides this, according to Plato's illustration, neither the poet nor the painter has true knowledge of what they represent as artists. Representation in art, Plato believes, could destroy knowledge. As a result, what Murdoch interprets here is 'the man who is painting the bed, where [Plato] indicates that the painter is pursuing some extraordinarily trivial activity, inferior even to that of the carpenter.'[38]

III. Plato rejected art for he 'thought of art as emotional fantasy. It gave expression to the lowest part of the mind of the artists and appealed to the lowest part of the mind of the client.'[39] This is where Plato directs his attacks especially on the contemporary tragedy as the most influential – on the audience as spectators, readers and hearers – of all artistic forms of the time. In the case of a representation, Murdoch believes that representation is substituted for reality; and this is where Plato thought of an emotional substitution.[40]

Besides these, what needs to be asserted is that the Romantic movement in literature (and the Romantic theory of literature as well), Murdoch argues, has nearly compensated for the damage Plato and his criticism inflicted on representative arts. She believes that the Romantics, very much like what Aristotle did, idealised art. The Romantic interpretation of art relies on the same view that art is an imitation. The difference, however, is the fact that although art is an imitation, it is the imitation of 'not particulars but the Ideas or Forms themselves.'[41] It was until the Romantic movement in English Literature that the definition of poetry – literature – had been limited to the doctrine that art was imitation. Sidney says, for example, that 'Poesy, therefore, is an art of imitation for so Aristotle termeth it in his word mimesis.'[42] Besides Sidney, it was Dryden who asserts talking particularly about the contemporary drama that 'play is to be thought the nearest imitation of nature.'[43] Furthermore,

37 Murdoch, 'Art is the Imitation of Nature,' 245.

38 Ibid.

39 Murdoch, 'Art is the Imitation of Nature,' 245–246.

40 Murdoch, 'Art is the Imitation of Nature,' 246.

41 Murdoch, 'Art is the Imitation of Nature,' 247.

42 Philip Sidney, *The Defense of Poesy*, (Boston: Ginn & Company, 1890), 9.

43 John Dryden, *An Essay of Dramatic Poesy*, Thomas Arnold (ed.), (Oxford: Clarendon P, 1918), 21.

Alexander Pope similarly suggests that the poet should 'First follow nature.'[44] In addition to these, 'Murdoch thinks that the Forms are general concepts which, when contemplated, enable us to see the true nature of the world.'[45] Therefore, the Romantic doctrine of art is that the poet – or the painter – is metaphorically elevated to the status of the carpenter whose act of imitation, as Plato insisted, is more meaningful for the carpenter imitates those ideas or forms.

This situation, furthermore, could be extended, as 'The philosopher Murdoch developed her ideas not just in dialogue with Jean-Paul Sartre, Immanuel Kant, and Plato but also with George Eliot and "innumerable novels,"'[46] to the notion that 'The artist, in Schopenhauer's view, elicits from nature forms of very great universality and intellectual clarity and condensed significance which are lurking there to be discovered.'[47] Moreover, Murdoch 'enjoyed some acquaintance with German and French post-Kantian philosophy, and took a particular interest in Sartre. She also familiarized herself with … Hegel, Schopenhauer, Nietzsche, and Heidiegger.'[48] Similarly the artist who principally imitates could still be associated, in Platonist terms, with a higher sense of imitation. Murdoch suggests, 'As Shelley put it, "Forms more real than living man, nurselings of immortality." Art is a higher truth and the artist is a kind of priest.'[49] Moreover, such idealisation of the notion of the artist (the poet) is reminiscent of how Sidney refers to the ancient Romans who defined the role of the poet as 'a diviner, foreseer, or prophet.'[50]

4 Conclusion

Murdoch was a very prolific Anglo-Irish novelist who published works of narrative fiction most of which have been critically acclaimed for their great variety of interest in a number of philosophical and moral questions. The reason why Murdoch mostly practised in her fiction philosophical and moral enquiries was

44 Alexander Pope, *Essay on Criticism*, John Churton Collins (ed.), (London: Macmillan and Co., Limited, 1896), 3.
45 Elizabeth Burns, 'Iris Murdoch and the Nature of Good,' *Religious Studies*, 33.3 (1997): 307.
46 Marije Altorf, 'After Cursing the Library: Iris Murdoch and the (In)visibility of Women in Philosophy,' *Hypatia*, 26.2 (2011): 396.
47 Murdoch, 'Art is the Imitation of Nature,' 247.
48 A.E. Denham, 'Envisioning the Good: Iris Murdoch's Moral Psychology,' *Modern Fiction Studies*, 47.3 (2001): 628.
49 Murdoch, 'Art is the Imitation of Nature,' 247.
50 Sidney, *The Defense of Poesy*, 5.

that she was, at the same time, a philosopher who particularly acknowledged herself as a Platonist. Murdoch's Platonism as an artist provided her with the firsthand experience of understanding of and elaborating on Plato's theories of the representative arts which were characteristically defined by the idea of imitation. As Murdoch was a novelist, she was, according to the Platonist definition of art, an imitative artist. Therefore, Murdoch illustrated and analysed not only in her fiction but also in her philosophical essays the connotations of the Platonist notion of imitation. Her artist protagonists, for example, demonstrate the fundamentals of the Platonist world such as the illusory nature of existence and the impracticability of figuring out the material presence 'as other characters in Miss Murdoch's world have been doing in various ways.'[51] Murdoch specially agreed with Plato and believed that art was an imitation. Therefore, the principal point of Murdoch's theoretical interpretations is the fundamental Platonist assumption that art is imitation. Murdoch, however, differs from Plato largely on the metaphor of the painter for she associates the art of painting with reproduction while Plato with transformation. The artist for Murdoch reproduces. In other words, MWurdoch believes that the artist, whose art is denounced as false and deceptive by Plato, recreates. Even realism in art, Murdoch claims, has apparently fallen short of providing a fully comprehensive example of the Platonist concept of imitation – which was originally based on the ideas of propaganda, distraction and fantasy – for she postulates that experimental art is itself imitative as well. It could be asserted that although Murdoch agreed with Plato on the imitative nature of representative arts, her interpretations of the idea of imitation considerably acknowledge Plato's hostility to art.

Bibliography

Altindis, Huseyin. 'The World Will Go On Living: Resistence to Eurocentric Epistemology in Linda Hogan's *Power*.' *Interactions: Ege Journal of British and American Studies*. 26.1–2 (2017): 15–28.

Altorf, Marije. 'After Cursing the Library: Iris Murdoch and the (In)visibility of Women in Philosophy.' *Hypatia*. 26.2 (2011): 384–402.

Antonaccio, Maria. 'Imagining the Good: Iris Murdoch's Godless Theology.' *The Annual of the Society of Christian Ethics*. 16 (1996): 223–242.

Aristotle. *Poetics*. Joe Sachs (trans.). Newburyport: Focus Publishing, 2006.

51 A.K. Weatherhead, 'Backgrounds with Figures in Iris Murdoch,' *Texas Studies in Literature and Language*, 10.4 (1969): 647.

Burns, Elizabeth. 'Iris Murdoch and the Nature of Good.' *Religious Studies*. 33.3 (1997): 303–313.
Denham, A.E. 'Envisioning the Good: Iris Murdoch's Moral Psychology.' *Modern Fiction Studies*. 47.3 (2001): 602–629.
Dryden, John. *An Essay of Dramatic Poesy*. Thomas Arnold (ed.). Oxford: Clarendon P, 1918.
Eagleton, Terry. *Marxism and Literary Criticism*. London: Taylor&Francis, 2006.
Murdoch, Iris. 'The Fire and the Sun: Why Plato Banished the Artists.' *Existentialists and Mystics: Writings on Philosophy and Literature*. New York: Penguin Books, 1999.
Murdoch, Iris. *The Black Prince*. New York: Penguin Classics, 2003.
Nicol, Bran. 'Iris Murdoch's Aesthetics of Masochism.' *Journal of Modern Literature*. 29.2 (2006): 148–165.
Plato. *The Republic: Cambridge Texts in the History of Political Thought*. G.R.F. Ferrari (ed.). Tom Griffith (trans.). Cambridge: Cambridge UP, 2000.
Pope, Alexander. *Essay on Criticism*. John Churton Collins (ed.). London: Macmillan and Co., Limited, 1896.
Sidney, Philip. *The Defense of Poesy*. Boston: Ginn & Company, 1890.
Sontag, Susan. 'Against Interpretation.' *Against Interpretation and Other Essays*. New York: Penguin Classics, 2009.
Sturrock, John. 'Reading Iris Murdoch.' *Salmagundi*. 80 (1988): 144–160.
Weatherhead, A.K. 'Backgrounds with Figures in Iris Murdoch.' *Texas Studies in Literature and Language*. 10.4 (1969): 635–648.

CHAPTER 9

Poetry as Place: the 'Vrai Lieu' in the Work of Yves Bonnefoy

Layla Roesler

Abstract

The notion of poetry as a 'vrai lieu', or 'true place', is a recurring trope in the work of French poet Yves Bonnefoy (1923–2016) and is strongly linked to his dual goal of transmitting the theo-philosophical experience of 'presence' while combating the deconstructionist suggestion that language can never attain the objects it seeks to describe. His spatialization of writing is built upon two tools derived from linguistics, namely, the notion of 'difference' and that of 'deixis'. Bonnefoy legitimizes his assertion that poetry can materially harbor presence through the structuralist notion of 'difference', which posits a metaphysical space created within the structure of language. Deixis, by contrast, allows him to postulate the sense of communion and physical sharing associated with the notion of presence. It is in this conjoined attention *that* Bonnefoy perceives presence as textually possible. And yet, this notion of space, of a 'true place' that would be poetry is, in fine, no more than a metaphor. Its frequent recurrence in Bonnefoy's work invites examination: what is the function of this metaphor? How does Bonnefoy use it to legitimate his programme?

Keywords

Yves Bonnefoy – French poetry – contemporary poetry – metaphor – place – 'vrai lieu'

•••

As a number of critics have noted, the notion of space, or place, is a central dynamic in the poetics of French poet Yves Bonnefoy. Alex Argyros, referring to Bonnefoy's first collection of poetry, *Du Mouvement et de l'immobilité de Douve*, writes: 'By far the dominant trope in the poem is that of place. From 'Théâtre', to the various 'vrai lieu''s to 'Orangerie', to 'Chapelle Brancacci', to 'His [sic] Est Locus Patriae', the poems which comprise Douve are repeatedly concerned

 | DOI:10.1163/9789004394315_010

with the question of a finite location'.[1] Marie-France Renard takes a similar perspective in her article, 'Les Structures Spatiales dans Du Mouvement et de l'Immobilité de Douve', where she underscores the importance of space for Bonnefoy, writing that 'The titles of his collections of poetry [...] as well as the themes developed there [...] or even the images themselves are equally imbued with space'.[2] The same theme, though at a symbolic level, is even more extensively developed in *Poétique du Lieu dans l'Oeuvre d'Yves Bonnefoy*, Patrick Née's very complete analysis of Bonnefoy's relation to the imagined places of Rome and Egypt.

These critical positions are built upon Bonnefoy's own stance, which is aptly illustrated by his frequent description of poetry as a 'vrai lieu', or 'true place'. This notion of 'true place' enters into resonance with his primary aim of conveying the theo-philosophical notion of 'presence' through his poetry. His argument is that poetry's particular use of language gives it the unique capacity to capture and transmit this sense of 'presence'. This it can do far more efficiently and authentically than all other forms of literature because of its reliance on the materiality of language. Over his long writing career, Bonnefoy increasingly became an adamant defender of the importance of the notion of presence, which he read as an intense reflexive experience of self and other that impacts not only the individual, but even society as a whole. This in turn led him to combat, at times ferociously, the deconstructionist suggestion that language is unable to render the objects it seeks to attain and cannot really ever account for presence. For Bonnefoy, on the contrary, the poetic language can, by the authentic quality of the relation it creates, bring into existence this 'true place' where presence can be transferred from the writer to the reader. As he writes in *L'Improbable*, 'Poetry happens within the space of utterance, but each step therein is borne out in a world reaffirmed',[3] confirming thus his ideal vision of a poetic space from which the reader might regain the world.

And yet, the spatiality of poetry has its own ambiguity, as Alex Argyros brilliantly shows in his analysis of the interaction between two poems from the Douve cycle. Argyros demonstrates that in this instance, Bonnefoy's 'true place', that is, the extra-linguistic place of real presence to which poetry would return is none other than poetry itself. The poem 'Hic est Locus Patriae' thematizes

---

1 Alex Argyros, 'The Topography of Presence: Bonnefoy and the Spatialization of Poetry', *Orbis Litterarum* 41 (1986): 253, emphasis in original.

2 Marie-France Renard, 'Les Structures Spatiales dans *Du Mouvement et de l'Immobilité de Douve*', in *Yves Bonnefoy: Poésie, Art et Pensée* by Yves-Alain Favre (Pau: Cahiers de l'Université de Pau et des Pays d'Ardour, 1986), 121.

3 Yves Bonnefoy, *L'Improbable* (Paris: Gallimard, 1980), 132.

this intricable loop within which language is caught by making the central character return to a place which is italicized and capitalized in the text, ('Ainsi le jour baissait sur le lieudit *Aux Arbres*'). This particular typography reveals it to be the title of the earlier poem, 'Aux Arbres'. 'In Douve, therefore', Argyros writes, 'a later poem refers to an earlier poem as a place. "Aux Arbres" is not simply a set of signs expending themselves in the service of a the pre- or post-semiotic, it is also, and perhaps foremost, a destination'.[4]

Clearly, what Bonnefoy has done here is materialize the 'placeness' of poetry via a sort of linguistic sleight of hand that blurs the referential boundaries of language. Hence, Argyros concludes that 'even the classic opposition between image and metaphor, or between writing as a transitive act and writing as an intransitive act, is rendered indistinct by Douve's retranscription of the notion of place'.[5] This is crucial, for even if we accept this use of language to create place, the fact remains that Bonnefoy's notion of poetry-as-place is itself a metaphor, and that point alone invites examination. This article, by showing that the metaphor itself plays a fundamental role in Bonnefoy's ambition to transmit the notion of presence, thus attempts to respond to the question of how this metaphor functions and why Bonnefoy relies so heavily on it.

Before addressing the function of the metaphor, some attention must be given to the nature of this 'vrai lieu' that is such a recurring trope for Bonnefoy, particularly in his earlier work. Indeed, no reader of Bonnefoy's first volume of poetry, *Du Mouvement et de l'Immobilité de Douve* can fail to recognize the importance that 'place' holds in his mental landscape. The climax of the five 'acts' that compose the volume is tellingly entitled 'Vrai Lieu'. Additionally, within that section three of the six poems bear the word 'lieu' in their title: 'Lieu du Combat', 'Lieu de la Salamandre', and 'Vrai Lieu du Cerf'. Regarding this attachment to the notion of place, Bonnefoy has made more general comments where he explicitly equates poetry with the true place, simultaneously asserting that the 'true place is always a here'.[6] Both these elements are vibrantly shown in the title of the poem mentioned above, 'Hic est Locus Patriae', which thematises the deictic notions of 'place' (*locus*) and of 'here' (*hic*). This notion of 'hereness' is vital to understanding the spatialization of text that is such a crucial element in Bonnefoy's work, appearing in his numerous references to the 'here and now': 'Poetry, like love, must decide that beings are. It must devote itself to this Here and this Now that Hegel proudly dismissed in the name of language'.[7]

4 Argyros, 'Topography of Presence', 257.

5 Ibid., 257.

6 Bonnefoy, *L'Improbable*, 121.

7 Ibid., 100.

This 'here' that Bonnefoy insists on has an underlying metaphorical function. And although the use of metaphor is perhaps a standard rhetorical device in poetry, the quality and function of the metaphor Bonnefoy is using stands in stark contrast to something like Éluard's 'La terre est bleue comme une orange' ('The earth is blue like an orange'),[8] whose poetic intent is to produce a 'living' metaphor to generate a reaction of shock. In what could almost be read as an inadvertent echo to Eluard's metaphor, Bonnefoy frequently evokes the physical space of the orangery. It is, for instance, the title and the space described in section four of *Douve*. From comments made in his essays, it is clear that he gives metaphorical portent this space, describing it in 'The Act and Place of Poetry' as the 'an index of the night' for its capacity to prefigure the openness of being.[9] But if these spatial metaphors do not have the same function as Eluard's 'poetic explosion', and if they are not the object of self-analysis for him, they are nonetheless not inert. On the contrary, and as has been convincingly argued by linguists George Lakoff and Mark Johnsen in their seminal work, Metaphors We Live By, even dead, metaphors are not mere rhetorical frills dusted ex post facto over language to embellish communication. On the contrary, they reflect the fact that 'human thought processes are largely metaphorical', meaning that we construct our life experience via representations that are essentially metaphorical, and these in turn condition our understanding of the conceptual systems that determine our vision of the world.[10]

Lakoff and Johnsen's insight into the broad-reaching impact of metaphor is based in part on Michael Reddy's earlier work on the so-called 'conduit' metaphor, and which he identifies as the dominant metaphor governing the way we think about the relationship between language and meaning.[11] As Reddy points out, the perception that language 'contains' meaning and that this language-container can be used to transfer meaning from one person to another seems so natural that it becomes invisible, making it difficult to see that the idea of words having a spatial volume with an 'inside' and 'outside' is no more than an image.

Bonnefoy's spatialization of poetry through his strong identification of it as the 'true place' correlates well with Reddy's conduit metaphor in the sense that positing the idea of a container as a space with an inside and an outside implies a differentiation that very naturally extends to the notion of a place within which something can be located. The naturalness of this metaphor

8 Paul Eluard, *L'Amour la poésie* (Paris: Gallimard, 1929), 18.

9 Bonnefoy, *L'Improbable*, 114.

10 George Lakoff and Mark Johnsen, *Metaphors We Live By* (Chicago: University of Chicago Press, 2003), 7.

11 Michael J. Reddy, 'The Conduit Metaphor-A Case of Frame Conflict in our Language about Language', 286–287.

is so complete that it is perfectly accepted by readers and Bonnefoy alike, as shown by the fact that Bonnefoy, despite the acuity he usually demonstrates on points of language and theory, never explicitly questions the metaphoricity of his 'vrai lieu'. Instead, he treats it as a wholly legitimate conjoining of language and location, even extending the metaphor to give, for example, a vision of a poetry that builds castles out of language to house the evanescent beings it hopes to harbour. The importance of this latter image is highlighted by the fact that it appears both in poetry ('I will name desert, that castle that you were'[12]) and in essay, where Bonnefoy, calls up the image of Dante creating a castle for Beatrice out of poetic words.[13]

The Lakoff and Johnsen theory is particularly efficient in explaining why Bonnefoy's metaphor seems so legitimate. Even if his initial choice was unconscious, the conception of poetry as place functions in a highly normative way because it relies on a spatialization metaphor and, as Lakoff and Johnsen point out, 'most of our fundamental concepts are organized in terms of one or more spatialization metaphors'.[14] This reinforces the fact that Bonnefoy's metaphor is coherent within a larger system of representations, which is exactly what Lakoff and Johnsen stress in their work, writing that: 'There are many possible physical and social bases for metaphor. Coherence within the overall system seems to be part of the reason why one is chosen and not another'.[15]

And yet, though they are clearly interconnected, the normative metaphor of language as a container into which meaning is inserted by the speaker/writer and then extracted by the hearer/reader is not identical to the metaphor of place. On the contrary, the notion of poetry as 'place' adds the spatial vector of a 'here' that is a crucial element of the deixis –literally, indexicals, or words that point—that is such a striking characteristic of Bonnefoy's work. It is thematised, for instance, in his statement that 'far from being a deforming prism that incites only to illusion, the poem is the index which points outside language'.[16] This 'outside' he reads as the real presence of things and beings.

Fundamentally, thus, the metaphor of poetry as place is strongly linked to Bonnefoy's explicitly stated goal of using poetry to give the reader a window to the experience of 'presence'. 'Poetry', he writes, 'does not exclude presence; it creates it'.[17] Indeed, the metaphor itself contributes to the goal. As Lakoff and Johnsen showed, the nature of metaphor is to create some possibilities by

12 Yves Bonnefoy, *Poèmes* (Paris, Gallimard, 1978), 73.
13 Bonnefoy, *L'Improbable*, 107.
14 Lakoff and Johnsen, *Metaphors*, 18.
15 Ibid., 19.
16 Yves Bonnefoy, *Entretiens sur la poésie: 1972–1990* (Paris: Mercure de France, 1990), 264.
17 Yves Bonnefoy. *Le Nuage Rouge*, (Paris: Mercure de France, 1992), 280.

logically extending the range of a concept and eliminate others that cannot be brought within its field.[18] Thus Bonnefoy's use of poetry as place is a logical extension of the language-as-container-of-meaning metaphor, while something radically different, such as poetry as fruit (to comically extend the Éluardian metaphor) is more difficult to accept.

Three possibilities play into Bonnefoy's desire to find a means to make language become a genuine receptacle for presence via the poetry-as-place metaphor. The first addresses the 'inside/outside' dichotomy that is created by the metaphor. The second relies on Bonnefoy's image of 'true-place-as-here'. This 'hereness' is the fundamental building block of the deixis that is so crucial to his work that he qualifies it as the very 'entryway to poetic creation'.[19] Deixis itself is complex because it refers to a conceptual category of language whose meaning is instable and always context-dependent. In brief, however, the deictic word is defined as one whose immediate referent can only be filled at the moment of enunciation. Thus the stable definition of the word 'here', understood as 'the word designating a location proximate to the speaker', varies with the speaker. This characteristic of obtaining complete meaning only within the individual speech act grounds Bonnefoy's argument for the status of presence in language, inasmuch as deixis points to the incontestable presence of its referent and presumes the reality of such a presence.

As we saw earlier, Reddy's 'language-as-conduit' metaphor is grounded on the idea that meaning can be placed within language and then transferred to others. This in turn leads to the vision of language as a container with an inside and an outside. In similar fashion, the notion of poetry as place implies the possibility that something can be inside this place. Clearly, that 'something' is, for Bonnefoy, presence itself. Indeed, other images that are frequent in his work reinforce this idea. Thus critic Jean-Pierre Richard, who has evoked Bonnefoy's recurrent metaphor of a 'snare', reads his poetry as an attempt to create a trap to hold presence.[20] By extension, the reader would then be able to find and extract the presence placed in the receptacle of poetry, thus confirming the usefulness of the poetry-as-place metaphor.

The outgrowth to this notion suggests that if presence can be trapped, then it can be forced to remain 'here', in the place designated by poetry. It is this metaphor of a 'hereness' (and its corollary, 'now'—the deictic *hic et nunc*) located in poetry that gives deixis as the solution to Bonnefoy's paradoxical

18 Ibid., 14.

19 Electronic note to me, May 2012.

20 Jean-Pierre Richard, *Onze Études sur la Poésie Moderne* (Paris: Éditions du Seuil, 1964), 221.

problem of how to indicate an extra-linguistic notion like presence through language. He bases his argument both on deixis's capacity to point to an incontestable referential reality and the fact that deictic words cannot be completely contained within a referential language system because they require the individual speech act to actualize them.

Thus, it is this 'deixis of here' that allows Bonnefoy to posit the third possibility, namely how language, through the Saussurian notion of 'parole' ('speech act', 'utterance'), can take on the shape of reality and thus incorporate presence. 'Parole', in a Saussurian sense, is the antithesis of 'language', which produces concepts rather than reality. All aspects of the reality truly experienced by individuals in their embodied, incarnated selves, are, for Bonnefoy, encapsulated in their use of 'parole'. And it is precisely the double capacity of 'parole' –both the expression of the individual speaking being and an accurate reflection of reality— that gives it the potential to harbour presence. As he writes: 'The difficulty of poetry is that language is a system, while the speech of poetry is presence'.[21]

But concretely, how can 'parole', the ordinary utterance of anyone, become presence? Bonnefoy's solution to the problem relies on 'parole' and deixis to provide a theoretical perspective that criticizes the deconstructionist rejection of language's capacity to express presence. Indeed, in response to Hegel's well-known observation that language cannot apprehend 'here' and 'now' as objective entities, but can only grasp them as absences,[22] Bonnefoy acknowledges the impact of the position, but indicates his visceral opposition, which is materialized in the capacities he gives to the 'true place' of poetry. In response to Hegel's conjoining of 'now' and 'night' to indicate language's incomplete grasp of reality ('To the question: "What is Now?", let us answer, e.g. "Now is Night"'),[23] Bonnefoy suggests that even if Hegel might have been randomly chosen his words, the semantic weight of 'night' translates an anguish to which poetry is the response:

> And yet, we have to move forward through the shadows. And if now is night, nothing says that tomorrow by the grace of a few poems, some light will not appear.[24]

21 Yves Bonnefoy, *Entretiens sur la poésie: 1972–1990* (Paris: Mercure de France, 1990), 97, emphasis in original text.

22 Georg W.F. Hegel, *Phenomenology of the Spirit,* trans. A.V. Miller (Delhi: Motilal Banarsidass, 1998), 60.

23 Ibid., 60.

24 Yves Bonnefoy, *Vérité de Parole et Autres Essais* (Paris: Gallimard, 1995), 63.

Bonnefoy's project is ambitious, for if language itself is the problem for Hegel's approach to the deictic hic et nunc, Bonnefoy suggests a means to resolve the problem by integrating deixis into language's ability to produce presence. Paradoxically, Bonnefoy's use of 'parole' itself relies on a form of deixis in order to resolve the problem of the incompatibility Hegel describes. Indeed, Bonnefoy accepts the Hegelian mistrust of language, stating that 'yes, it's true, the speech act empties the here and now of its concrete content, of its breadth of existence insofar as it attempts to describe these things.[25] And yet, he continues, 'It still has the possibility of *designating* this here-now, if it breaks with the idea of describing it'.[26] The importance of 'designating' cannot be gainsaid; Bonnefoy points to it by italicizing it. In so doing, he materializes the concrete function of 'designate', which etymologically denotes marking 'with a sign' to bring something to the attention of others. 'To designate' thus means to point to something, and this is, of course, the very function of deixis. In opposition to Hegel, thus, Bonnefoy clearly suggests a rehabilitation of language through the symbolic pointing of the italics and in the semantics of the word 'designate'. The argument can appear surprising because of its seeming circularity: the inability of language to indicate the objects of deixis can be undone by deixis itself. However, this position also confirms the double function of language, which is always both descriptive and indicative. Bonnefoy's 'désigner' would thus point again toward parole as the only agent of language able to undo the system to which it belongs in order to allow for the expression of presence.

Finally, deixis allows Bonnefoy to postulate the sense of communion and physical sharing associated with the notion of presence. The textual 'here' which writers use to refer to their own work is a deictic that uses words as symbols pointing not extra-linguistically, but rather toward the space within the text. But this imaginary space, this 'here' for the writer, becomes, through the act of reading, a 'here' for the reader as well because of the conjoined attention that both writer and reader give to the physical space of the written text. It is in this conjoined attention that Bonnefoy perceives presence as textually possible. His metaphor of place expands, however, the possibility of integrating presence into a three-dimensional space that stands in opposition to the ordinary limits of writing. Indeed, though the written text quite literally creates a physical space which is concretely inhabited by symbols also of a physical nature (e.g. shape, style, typography, etc.), it remains bound to the two dimensions that are its existential lot.

---

25 Yves Bonnefoy, *Yves Bonnefoy: Poésie, Art et Pensée*, by Yves-Alain Favre (Pau: Cahiers de Université de Pau, 1983), 45.

26 Ibid., emphasis in original.

The metaphor of place thus functions in Bonnefoy's work by expanding these limits through the creation of a form more consonant with the reality of the material world. Consequently, it provides this poetry-as-place with a greater capacity to integrate the deictic 'parole', and through it, presence. To extend this image just a bit further, one can say that only real things can enter a three-dimensional space. It is clear that for Bonnefoy, nothing is more real than the notion of presence. The metaphor of the 'vrai lieu' thus exists to create a place of presence and consolidate the reality of this textual presence.

Bibliography

Argyros, Alex. 'The Topography of Presence: Bonnefoy and the Spatialization of Poetry'. *Orbis Litterarum* 41 (1986): 244–264.

Bonnefoy, Yves. *Entretiens sur la poésie: 1972–1990.* Paris: Mercure de France, 1990.

Bonnefoy, Yves. *L'Improbable.* Paris: Gallimard, 1980.

Bonnefoy, Yves, *Poèmes*. Paris: Gallimard, 1978.

Bonnefoy, Yves. *Vérité de Parole et Autres Essais*. Paris: Gallimard, 1995.

Eluard, Paul. *L'Amour la poésie.* Paris: Gallimard, 1929.

Favre, Yves-Alain, ed. *Yves Bonnefoy: Poésie, Art et Pensée*. Pau: Cahiers de Université de Pau, 1983.

Hegel, Georg W.F. *Phenomenology of the Spirit.* Translated by A.V. Miller. Delhi: Motilal Banarsidass, 1998.

Lakoff, George and Mark Johnsen. *Metaphors We Live By.* Chicago: University of Chicago Press, 2003.

Renard, Marie-France. 'Les Structures Spatiales dans *Du Mouvement et de l'Immobilité de Douve*'. In *Yves Bonnefoy: Poésie, Art et Pensée*, edited by Yves-Alain Favre, 118–132. Pau: Cahiers de l'Université de Pau et des Pays d'Ardour, 1986.

Reddy, Michael J. 'The Conduit Metaphor-A Case of Frame Conflict in our Language about Language', in A. Ortony, *Metaphor and Thought,* Cambridge: Cambridge University Press, 1992, 284–310.

Richard, Jean-Pierre. *Onze Études sur la Poésie Moderne*. Paris: Éditions du Seuil, 1964.

CHAPTER 10

Timeless and Spaceless Writer: the Case of Bediuzzaman Said Nursi

Yadigar Sanli

Abstract

Bediuzzaman Said Nursi (1878–1960) has his place amongst the leading Islamic scholars in Turkish history. His written record spans in excess of 6000 pages in his 82 year life-time. His writings were nothing short of revolutionary, even sparking fear in the ruling authorities of his time. From 1926 until his death in 1960 his life was spent entirely either in prison or in exile. During his exile, he was under close observation and was stripped of much of his freedom to do or go as he wished. It was during these testing years, devoid of access to a single reference book, that the majority of his works were written. This is much like the French writer Fernand Braudel, whose defining work materialised whilst detained under Nazi concentration camp. Bediuzzaman Said Nursi's complete works titled Risale-I Nur (The Works of The Light) feature an interpretation of Qoran and Hadiths with contemporary logical themes and notes. These philosophical masterpieces came to light while he was imprisoned or exiled to the mountain villages. Said Nursi's writings, whilst accepted as academic works, can be simultaneously classed as fiction, owing to their literary construction. Such is his success in writing technique and style that his work is both accessible to, and easily understandable by, both educated and uneducated readers alike. In this contribution, I will question the context in which space-time-material is important to a writer, with due regard to publications requiring deep analysis or cross-referencing from multiple sources. To illustrate the discussion, I will be drawing on examples from the life of Said Nursi and his philosophical and fictional writing.

Keywords

Bediuzzaman – Said Nursi – risale-I nur – revolutionary works – timeless writer – philosophical writings

•••

 | DOI:10.1163/9789004394315_011

1 Introduction

One of the Islamic worlds' well-known scholars, Bediuzzaman Said Nursî, started life in 1878 in Nurs village near the Hizan district of Bitlis, which is currently Eastern Turkey. His 'Nursi' name originates from the name of his birthplace. As he was part of the Kurdish community, he was also known as 'Said-i Kurdi' (Said the Kurd).

As well as Islamic studies, he also completed an array of positive science subjects at the age of fifteen. These included geography, mathematics, physics, chemistry, and the like. He was to excel in these various branches of science and knowledge and honoured with the title of 'Bediüzzaman', meaning 'unique of the time'. He passed away in 1960.

2 A Very Peaceful Life

The life of Said Nursi after 1926 passed both in exile and in prison as a result of the books he wrote.

One of the reasons he was imprisoned/exiled was his choice to defend the freedom and rights of individuals, within the general public as well as every single creature in nature, against injustice. The ruling despotic regime minimised his freedom because of his Islamic belief but could never stop him fighting for his ideology, writing books and training students. Because, he had a mission: to make the people understand the real freedom imparted by Islam. Bediüzzaman's description of 'Freedom' being: an individual living without harming himself or any other. In his struggle to defend freedom, he believed that no reason could justify harming others. This was his understanding of freedom according to The Quran.[1]

Nursî stood in 'opposition' to enmity. The truths he declared in every occasion were to avoid each and every action which would make the world a worse place to live. He taught the avoidance of hatred which prevents the weak and poor reaching freedom and fairness. Therefore, Nursî expressed 'Love' but avoided 'hatred', uttered 'fairness and honesty'. Bediüzzaman was an Islamic scholar who spoke out of a philosophy of being 'Free'. His faith lessons taught against mayhem and terror.

Bediüzzaman recognised that the main problem facing humanity was ignorance/illiteracy for. He also expressed that 'Dominance' will happen via knowledge not with the sword. Throughout his entire life, Bediüzzaman

---

1 Said Nursi, *Tarihçe-i Hayat* (Istanbul: Söz Basım Yayım, 2006), 106.

declared 'ignorance, poverty and discrimination' as enemies to himself. He recommended that, against those three enemies, 'knowledge, art and unity' were the appropriate weapons with which to arm oneself. Bediüzzaman campaigned not against humanity but against those who harm society with 'Love for Love, hatred for hatred'. That was Said Nursi's slogan.

He greatly disliked politics. He never read a newspaper nor paid attention to a radio broadcast and never asked anyone to relate any political news after the age of 40. He even strongly prohibited his students from politics. He was not a man of politics. According to him, politics was concealing, and throwing dirt on the fight for humanity and the entire being.[2] 'Euzu billahi mineşşeytani vessiyase' meaning:

> I seek refuge in Allah against satan and politics' was another of his slogans. 'We as students of Risale-i Nur, will never make Risale-i Nur a tool, not for worldly issues, even universal issues. And also, Quran has strongly prohibited us from Politics.[3]

'Kindliness, Rightness and Truth as the main objectives of the Risale-i Nur, strongly prohibited us from politics and standing against the government'.[4] The fact that he didn't like politics was never a justification to stand against the government. Although he was tortured by the ruling powers, but never stood against them. Because he always believed that whilst humans have the potential to make mistakes at a governmental level, the most important thing is keeping the country unified in peace.

3 Forbidden Writer-Forbidden Books: Exile-Prison, Exile-Prison, Exile-Prison

In 1926, Bediüzzaman was deported to western Anatolia from the east where he was born, grew up, and where he dreamed to set up a university. He never returned back home. Because, the person who defended freedom all his life, never had a chance to choose where to go. Deportation and total isolation continued in force for 35 years until the end of his life in 1960.

Bediüzzaman's life and his works after 1926, show a similar pattern to French writer and academic Fernand Braudel. When Braudel was held at a

2 Said Nursi, *Asa-yi Musa* (Istanbul: Envar Nesriyat, 1994), 20.
3 Said Nursi, *Tarihçe-i Hayat*, 691.
4 Ibid., 691.

Nazi concentration camp, he created a masterpiece without books or a library. From this perspective, Bediuzzaman and Braudel are both very rare figures who created works of mastery under extreme conditions, without referring to any external source.

Over 6000 pages, 130 books and letters titled as Risale-i Nur (Books of the Light), were taken under inspection with the reason that religion has become a tool for politics. Books were forbidden, even writer Nursi was charged with the death sentence by the secular bureaucrats who considered Islam an enemy to their political ruling system. However, all these accusations were acquitted by the court of law: Bediüzzaman's crime was to write books. When he started to write during exile in Burdur province, he was sent off (exiled) to a small mountain village, Barla near Isparta where transport was not available.[5] The intention was to keep him away from his students. There were no university educated students around him in this mountain village. But, this time God gave him illiterate students. The (illiterate) villagers looked after him and paid attention to his works. Hundreds of villagers learned to read and write as a result of these works.

It was a real success that Bediüzzaman's works were published and distributed in a time that, all religious scholars had been silenced and their works barred.[6]

The ill-treatment was not limited to Bediüzzaman, but people who approached him to ask a question in order to further learning, were also prosecuted and imprisoned.[7]

People who copied Risale-i Nur (The Books of the Light) by hand in their own homes were arrested, imprisoned and tortured, and houses searched repeatedly by soldiers.

Bediüzzaman's relatives were also banned from visiting him during isolation and exile.[8]

4 Nursi's Writing Conditions and Materials

Why did he want to write? What were the reasons that drove him to write? According to Said Nursi, the time and focus of man now, is not in personal issues, but what is of utmost importance is the masterpiece. His Risale books

5 Ibid., 192.

6 Ibid., 193.

7 Ibid., 203.

8 Said Nursi, *Mektubat* (İstanbul: Söz Basım Yayım, 2006), 78.

redirected minds to the actual, original Work of Divine Mastery – The Quran. This Mastery materialises in nature as all that which is created and living. His main goal was to reach the truth through giving multiple examples within creation and asking questions. Reaching the truth itself is approached through a process of parts to whole, by questioning mind and conscience and also, convincing mind and heart.

Why was he so insistent in writing in spite of all the prevention? Braudel has stated that: civilisations are a result of continuity of continuity, never ending historical continuity.[9] One of the reasons that drove Nursi to write was to salvage civilisation from breaking off. Particularly, this can be seen in the period after 1928. Maintaining continuity has been achieved partly by Nursi's work, committed under extreme conditions.

As İbni Câhız stated, 'Book will make death speak, and will deliver words of life'.[10] Nursî left his words for future generations by writing books, in spite of lack of time and materials. Nursi realized the importance of leaving something written behind, same as the Quran, that was also a written book. Words fade away but scripture remain. Therefore, even if it is at the cost of a life, he had to write. For the sake of writing, he returned from gallows tree and had been tortured in numerous ways. He was banned from pen and paper. He was poisoned. His compulsory residence was under surveillance constantly. Having window curtains was not allowed, so they could monitor his movements, and bar him from any writing. Despite this, nothing prevented him from writing. Neither timelessness, nor spacelessness

Publishing and printing Bediüzzaman's books were also banned by authorities. When he was in exile in the far corners of the country, villagers protected his works. In an extraordinary act of inspiration, peasants learned reading and writing in order to contribute to making copies of the books. In 1928 the Arabic alphabet was lifted by the new republic administration and the new Latin alphabet took over. Legally, using old Arabic letters was totally forbidden by law. The alphabet was compulsory. Many writers, poets, scholars were made illiterate overnight as a result of this new law. The old books had to be translated and converted to the new alphabet. In the Ottoman education system, the taught alphabet was Arabic. Nursi was also educated in this system, and naturally the Arabic alphabet was all he knew. In addition to that, the aforementioned

9 Fernand Braudel, *Uygarlıkların Grameri,* trans. Mehmet Ali Kılıçbay (İstanbul: İmge Kitabevi, 2014), 67.

10 Johannes Perdersen, *İslam Dünyasında Kitabın Tarihi,* trans. Mustafa Macit Karagözoğlu (İstanbul: Klasik, 2013), 5.

peasants were also familiar with the Arabic letters and only because of the Risal-i Nurs (The Books of the Light).

Risale books were banned from publication so they were multiplied by hand over a period of twenty years in secret. However, they had to wait another 35 years to be printed by publishers legally.[11]

Why are these Risale books so special? Of course, it is not down to the quality of paper, book cover or art of book making. Because these works were written on whatever material was available during their 25 year writing duration. He was a man that could read and memorize a few books overnight. His strong memory did not hold him back from writing books. Because the knowledge of the past also needs to be revised or amended as per the demands of the present. And he has done it. He used whatever material was available to write on even though he was banned from writing. But knowledge had to be passed on to future generations, with a new look and interpretation. At the first opportunity, these works were transferred to paper. Having difficulties, lack of time, material and availability was not a barrier to deliver 'Peacefulness' as a 'Free Man'.

Bediüzzaman wrote his first 384 paged book of Quran commentary 'İşârâtü'l-İ'câz' in 1915 when he was fighting against the Russians to protect his homeland, women and children of Van city. Under these conditions, on the war field, he had no reference book with him. He wrote the book, sometimes on horseback, during fighting or at a rear-site of the war.[12]

He wrote a 227-paged 'First Book of Nur' in Burdur city in 1926 when he was in exile. This book had been copied by hand and distributed in secret.[13] After that, he was taken to a remote location in Barla village of Isparta, in March 1927. Barla was somewhere very difficult to reach. The easiest way to get there was via lake by a boat. Barla was a perfect place to hide him, away from public exposure. His usual travelling pack was standard -a cane basket, pair of underwear, tea kettle and his flask to keep cold water. No pen or paper. The only book he had was a hand written old Quran. Later, this old Quran also became an excuse for an accusation of crime, and was taken away from him by the law courts. But, ironically, people of Barla took very good care of him and took his work very seriously. Barla became the first school of Nurs. The main books of the Risale-i Nur collection were written and multiplied by hand in Barla under very strenuous conditions. Nursi had access to only one person in Barla village that could read and write. In the ensuing months and years, gardens and fields

11 Said Nursi, *Tarihçe-i Hayat*, 205.

12 Said Nursi, *İşârâtü'l-İ'câz* (İstanbul: Söz Basım Yayım, 2006), 17; 24–25.

13 Said Nursi, *Tarihçe-i Hayat*, 191.

of Barla became classrooms of Risale-i Nur. Initially, people who could not write just copied pages on illuminated tables. Later, almost an entire village could read and write. As a meticulous academician, Nursi consistently went through every single hand written page and edited and corrected mistakes individually. His working space was countryside, mountain peaks, orchards and farms. Generally he chose quiet places. He ate a bowl of soup a day if available and always paid for this. He never accepted donations or charity from anyone.[14]

Keeping Nursi away from the public proved not to be an effective solution for officials. Somehow, Nursi managed to distribute his works and trained many students. Consequently, Nursi was consigned to life in prison. He was charged with no reason and put in prison in Eskisehir, near Ankara. He spent 11 months in prison together with his students, after which time he was discharged because of lack of evidence. But, it was not enough. He was this time taken to Kastamonu, another quiet location. Again, he was not allowed to write and distribute. He could not use national postal services to distribute his writings. Therefore, his students formed an unofficial voluntary basis volunteer postal network called 'Nur Postals'. Those volunteer postmen were farmers, bakers and shepherds. They distributed pages in a self-made system to others. Finding suitable paper was not possible everytime. Therefore, sometimes, they used flat breads and cigarette covers. Some of these materials have been preserved well by Nursi's living students and can be seen today.

These pages have been gathered following circulation and compiled into a book, known today as 'The Letters of the Book of the Light'.

In 1943, Nursi wrote a 169-paged book called 'Letters of Fruit' between two Fridays while he was imprisoned in Denizli city. Again, due to the absence of paper, this book was first written on matchboxes.[15]

The entire Risale collection was written against the odds, under extreme circumstances. A 188-paged '19th letter' was written within 12 hours as a result of working three hours a day, without using any source or reference book. Part of this work was written under the rain on the mountain. 34-paged '30th Word' which is philosophically heavy, was written in 6 hours in a fruit orchard. Another work of ten pages, called '28th Word' which explains difficult and philosophical matters about heaven, with a clear language to convince minds, was written in two hours in another garden.[16]

14 Ibid., 206.

15 Ibid., 493.

16 Said Nursi, *Mektubat*, 526–527.

Said Nursî wrote the 34-paged '30th Light', the 48-paged 'First 'Light' and the 46-paged 'Second Light' in Eskişehir prison, together with his 120 students, under very unpleasant and difficult conditions.[17]

Said Nursî was sent off to Afyon prison, together with another of his 50 students in 1948. He was deliberately kept in an ice cold prison cell in winter time and poisoned. During this time, he was already an old and sick man. He survived and carried on his mission. He kept writing; His students managed to send written pages out of prison.[18]

Nursi's books were copied by hand and distributed to the wider public for 20 years in secret until 1947. Approximatley, 600 thousand hand written works had been distributed until employing a printing machine.

Every single printed copy was habitually sent to Nursi for his approval. Every time, tirelessely, Nursi would oversee the pages and approve them.[19]

During all these difficult years, while Nursi was writing his books, he did not have any reference book with him apart from the Holy Quran.[20]

Bediüzzaman's life was far from easy. Despite this, he never made a single complaint about the mission he felt burdened to deliver, and never made an excuse to give up on writing. He firmly believed that the human mind should be freed from all bounds but Allah himself. He was not totally in agreement with the traditional methodology of understanding and expressing The Holy Quran and Islam. Often, he spoke out about his way of understanding. In his early years, he joined in many discussions between scholars and he clearly, bravely and strongly spoke out his thoughts. Some said he was crazy and should be taken to a mental institution, while others said he was extraordinary and unique – 'Bediuzzaman'. Therefore he was also named as a 'Free Man – Hur Adam'.

5 Conclusion

Bediüzzaman's works are deep, philosophical studies, calling for analytical thinking and reflection. Said Nursî set out to make the Holy Quran more accessible to modern man, opening up meanings and exploring understanding by using methods of The Quran itself.

17 Said Nursi, *Tarihçe-i Hayat*, 270.
18 Ibid., 677–679.
19 Ibid., 202.
20 Ibid., 203.

Said-i Nursi's works directly target the human mind. For a reader of Nursi, it is vital that the questioning mechanism of the mind should not be closed off. His works encourage the reader to ask questions. His writing method is based on questions and answers. All answers are eloquently supported and proven with current living and relevant examples. Therefore, in Nursi's works first the mind is to be convinced then the heart to be satisfied.

He neither had a comfortable place nor convenient time to work on his books. Prisons became madrasa/school for him and his students. But he always stood up for peaceful unity. From this perspective, Nursi's work undoubtedly deserves special attention and appreciation.

Bibliograpy

Braudel, Fernand. *Uygarlıkların Grameri.* Translated by Mehmet Ali Kılıçbay. İstanbul: İmge Kitabevi, 2014.

Pedersen, Johannes. *İslam Dünyasında Kitabın Tarihi.* Translated by Mustafa Macit Karagözoğlu. İstanbul: Klasik, 2013.

Bediüzzaman, Said Nursi. *Sözler*. İstanbul: Söz Basım Yayım, 2006a.

Bediüzzaman, Said Nursi. *Mektubat*. İstanbul: Söz Basım Yayım, 2006b.

Bediüzzaman, Said Nursi. *Asa-yı Musa*. İstanbul: Söz Basım Yayım, 2006c.

Bediüzzaman, Said Nursi. *Muhakemat*. İstanbul: Söz Basım Yayım, 2006d.

Bediüzzaman, Said Nursi. *Tarihçe-i Hayat*. İstanbul: Söz Basım Yayım, 2006e.

Bediüzzaman, Said Nursi. *Şualar. İstanbul*: Söz Basım Yayım, 2006f.

Bediüzzaman, Said Nursi. *Emirdağ Lahikası*. İstanbul: Söz Basım Yayım, 2006g.

Bediüzzaman, Said Nursi. *Kastamonu Lahikası*. İstanbul: Söz Basım Yayım, 2006h.

Bediüzzaman, Said Nursi. *Barla Lahikası.* İstanbul: Söz Basım Yayım, 2006i.

Bediüzzaman, Said Nursi. *İşaratü'l-İ'caz*. İstanbul: Söz Basım Yayım, 2006j.

PART 3

Meeting the Artist

CHAPTER 11

In-Movement: Spatiotextual Inscriptions

Panagiota Mavridou

Abstract

In-Movement derives from an impulse to bring together personal interests as diverse as writing, translation and the mystery of language(s), dance and performing arts, architecture and the city, as well as ongoing academic research on the embodied perception of place and aspects of space in literature. It draws upon the author's experience of writing in order to reflect on the way movement builds connections between writing and spaces. This experience involves composing mental texts while walking (in movement), which are afterwards distilled into written texts (in-movement). Essentially, a question poses itself: could movement emerge as a tool for investigation both on writing and spaces? *In-Movement* constitutes an open-ended experiment designed to explore, register, expose and comprehend the creative process(es) of writing in relation to spaces. The proposal is deployed in three stages, leading up to and building forth on the conference. During the first, rather ludic phase, primary material is created; for 10 consecutive weeks, one walk a week gives birth to a text, hence *A walk, a text, a week*. In parallel, a detailed monitoring of this exercise takes place, mapping and registering relevant spatial, temporal, psychological and physical information. The second, contemplative phase, implies revising and reflecting upon the material and the process. A draft paper is developed in the form of a *Post-Scriptum* to the spatiotextual dérives, addressing issues like: How does the writing performed in a space modify this very space? How are spaces inscribed in writing? Is this still-to-be-written text I compose inside me already a text? Where in my body does writing occur? The third phase, *Exposure*, opens the experiment up to the conference participants. Any dialogue, reaction and discussion produced during or after the conference, shall keep this project moving and exploring anew.

Keywords

Body – movement – walking – writing – spaces – inscription – experiment – performative

•••

 | DOI:10.1163/9789004394315_012

In-Movement was intended and designed to be an open-ended experiment, an experiential exploration of writing, the self, the body, movement, rhythm, spaces and places. Therefore, this paper attempts a reflection on the experiment, with focus and attention being directed towards the multiple elements and processes involved in writing throughout the duration of the project. In this *Post-Scriptum* to *A walk, a text, a week,* some of the doubts, dilemmas, struggles, and problems faced are registered and exposed a posteriori, opening up the discussion and inviting to further dwelling on the questions raised.

The departure point for this project was the simple observation that I often write while I walk. In movement, I find myself composing mental texts, writing in my imagination, or else, listening to the voice inside me pronounce those words; words which are afterwards remembered, recalled and written, into actual texts. It is with curiosity that I observe this process, trying to figure out how writing is layered in space and in time, but also what triggers it in the first place, where in the body it begins and what might be the external information that stimulates it.

While writing and walking, it becomes evident that writing is, essentially, a somatic act, a body movement as any other. The pace of the steps is connected to the pace of the heartbeat, the breaths taken, whereas those breaths are in turn connected to the rhythm and musicality of the words pronounced inside, to the heartbeat of the text created. Also, spaces and the written text become interwoven; they are mutually inscribed into each other. It should be fascinating to unfold the mechanisms of those inscriptions and look into them.

Hence the idea, quite straightforward, to monitor a personal and spontaneous experience and explore movement as a tool for investigation both on writing and spaces. What could go wrong?

The process was full of accidents, unexpected and unpredictable events, ranging from physical injury to multiple academic, professional and personal life changes, increasing in number and intensity towards the end. Naturally, all those changes have been incorporated and reflected into the project, given that such an experiment could not be immune to them. Thus, attention was partially redirected to the way they influenced the process and some unexpected observations emerged, introducing new aspects or issues.

A map of twenty-eight weeks was created as a unique temporal map, a walking-writing diary which documents the experiment and maps a person and a process (or maybe a person as a process) through time. This way, time emerged as another equally important element, while space and place remained inscribed in the walks, the texts and the reflections referenced in the map. Each page of the multifold diary represents a week of the project, whereas each snapshot on the back side stands for a week.

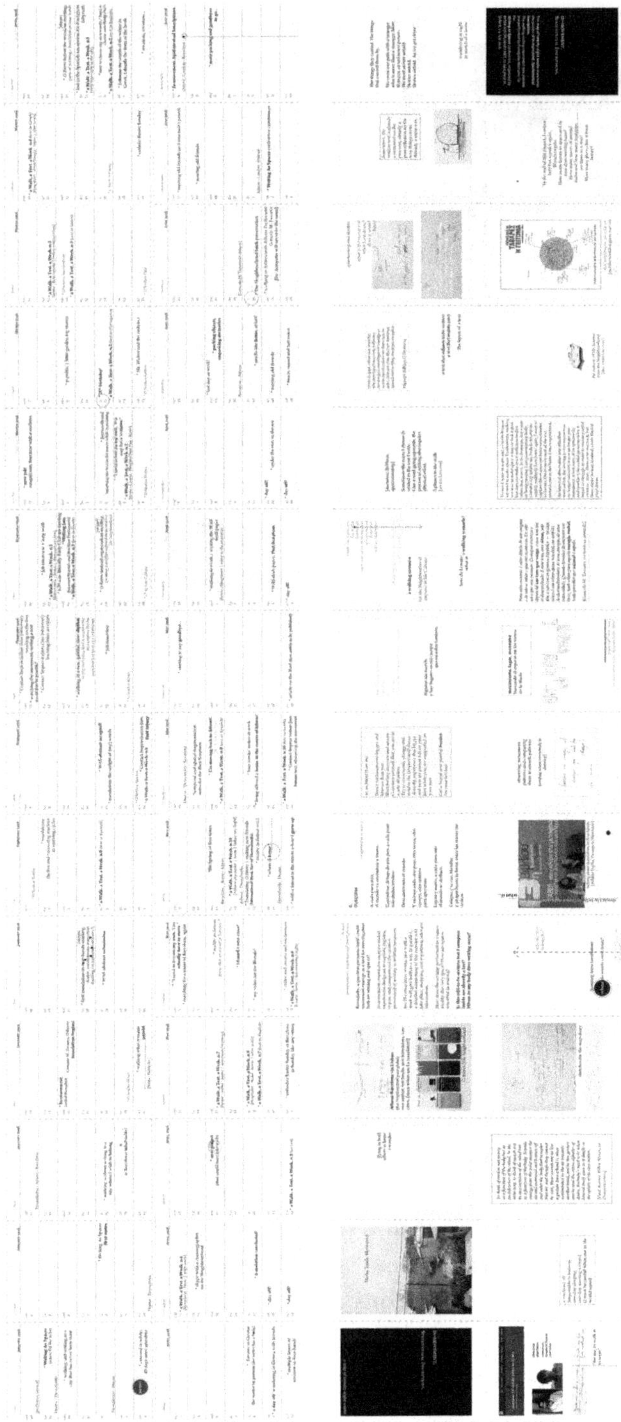

IMAGES 11.1 AND 11.2 'A map of twenty-eight weeks' unfolded: front side and back side. Original dimensions: 10,5 × 207,17 cm, folded into 10,5 × 7,2 cm

IMAGE 11.3 Aspects of 'A map of twenty-eight weeks'

To begin with, on the day I had planned to perform the very first walk of the series *A walk, a text, a week,* during a contact improvisation jam session, a colleague accidentally fell on my foot, causing a permanent injury and also a significant inflammation. A limp and the respective physical pain accompanied me for the next two months. It suddenly struck me: we take our bodies, their resistance, presence and reliability for granted. All of a sudden, I was unable to walk long distances and I was unable to walk fast; I had to reschedule all my activities, I had to adapt. The simple act of wandering in the city that I always cherished so much, became a painful and exhausting activity. During a walk, I couldn't focus on anything other than the pain, or the annoying feeling that made me want to burst into tears; gradually, ignoring the pain, breathing the new, slower pace and going back to observing people, trees, buildings, objects, or even my own thoughts and memories, became an almost meditation-like activity.

An accident happened and came to be a symptom that revealed a whole new world. Body parts and functions usually left in oblivion suddenly manifested themselves; they were no longer absent. They became present and this presence was reflected in my writing, a writing of the body and our experience of it and through it.

Just after completing the series *A walk, a text, a week*, I made the decision to move back from Barcelona to Athens. While revising the material for the paper, I was surprised to discover that this decision is clearly expressed in one of the texts written in movement. I feel that this fact contributes to the general idea that, when writing in movement, the text created might be a step forward in relation to deliberated or premeditated writing. In the same way that during a dérive my perception and intention often follow my steps (I suddenly find myself in a place), during a spatiotextual dérive my conscious thoughts might follow the ones inscribed in the text I create, semi-consciously (I suddenly discover a place in myself - a decision - where I didn't know I had been before).

During this intensely rich period, in accidents, new experiences and opportunities, I realized I was lacking space; temporal space, personal space, mental, psychological, but also physical space, in order to process, digest and react to what life was bringing to me. Everything felt fragmented and asphyxiated. In relation to writing, this lack of space generated a double, partially conflicting condition. On the one hand, at times it was almost impossible for me to write, either in movement or in paper; I was missing the essential hideout and shelter from the world. Then again, at times writing was my only shelter; performing long night walks and writing in the most unpredictable places - outside home, since home was falling apart - became my only way to a temporary home.

I was unaware of it, but the project gradually shifted towards an idea of what home means and what it is to inhabit. A strong link between writing and space was revealed. We need space to write and we write because we need to make space. Furthermore, walking is a way to make space when we lack it (just like dancers make space with their bodies when they move). In the footsteps, lies a way of being present. I am present in my body, I inhabit it (I inhabit my body to inhabit the world), suddenly I am home again. I want to capture this evanescent home in movement, so I write. The in-movement of the text reminds me of the home I once experienced. Could this personal speculation contribute to an understanding of the existential roots of writing?

In the beginning of this project, I was seriously concerned with the language in which it should be elaborated. I am a Greek woman living and studying in Spain for the last three years, with a special interest in Portuguese culture and language, which means that I alternatively and selectively express myself in three languages. I am also an architect fascinated by the relationships between body, language, space and narrative, so I find transitions from one language to another particularly exciting and revealing, often comparing them to nomadic ways of inhabiting the world. My recent experience with literary translation from Portuguese to Greek begins to add viewpoints to the discussion and shed light to certain aspects of the relationship between language and body structures.

However, this project should speak English, a language I studied during many years but never actually used much; it remains a stranger to me. In movement, it would be impossible for English words to be written, in any way, so I was worried that my proposal might be incommunicable, impossible to share in the conference. I therefore established a comforting rule. I would register and monitor all walks performed, but also take note of ideas and reflections related to the project, exclusively in English. Every meta-text would be originally written in English, just as this *post-scriptum*. On the other hand, I would let my texts in-movement emerge, the way they always did; in case I needed to share them in the conference, I would kindly ask a friend to help me with the translation.

Later in the process, I discovered that each of my texts selected to pronounce itself in its native language, building a multilingual writing condition. Texts have been written in Spanish, Portuguese, Greek and, surprisingly, English. It would be interesting to further explore this situation. Where is my language? How and why do I occasionally choose to inhabit one language or another? What part of me makes this choice? How does my style, attitude and writing change from a language to another? What remains the same? What is the nature of our connections and affinities with languages?

A walk, a text, a week also introduced another question, which arose as unexpectedly as an accident. In the decision to perform and at the same time monitor an experiment, dwells an essentially methodological problem, the problem of being simultaneously inside and outside the frame, subject and object of study. In a performative sense, being the product of one's own process. Walking the thin line in between is like experiencing a constant duality; for me, it meant constantly questioning my position, my motives, the sincerity and integrity of my actions. Wasn't I crossing delicate boundaries? Wasn't I betraying something?

Composing a reflective piece on the way things did or didn't work out as planned became of primary importance to me. My intentionality was directed towards the process of this experiment (which was performed in order to explore the process of my writing). Was I distancing myself from the object of study or getting closer to it? Was I walking in concentric circles? As the project advanced, it became clear that the experience of writing in movement and the observation of this experience are bound together. And as I was becoming more conscious of the process, this bond became stronger and I had difficulty separating the text from the meta-text, the writing from the reflection on the writing. In some cases, I caught myself composing meta-texts while walking, while the text was absent, inexistent. Or was I already composing the text of this *post-scriptum*?

Carrying out this project also changed the initial experience that inspired it. As layers of observation and reflection were added, the writing in-movement was no longer innocent; it was no longer pure experience, it lost some of its spontaneity and sparkle. Like a child growing older, it begun to struggle in order to remain playful, intuitive, pure. There is no way back, but there might be way forward, a new kind of playfulness. Or so I hope.

This last point is related to another problem inherent in the nature of the project: how can I let happen something that I am actually triggering? When I decide to perform *A walk, a text, a week*, can this walk really be intuitive, since it is by definition premeditated (I make space in my week, to make sure that it can happen)? After being aware of this contradictory situation, I responded to it by opening more space for the writing in movement to occur. I walked many times a week, almost daily, sometimes twice a day, all kinds of walks: functional and non-functional, with or without destination, in the morning, afternoon, evening or night, tired or after a nap, hungry or full, in a hurry, anxious, calm, before or after work, close to a deadline or right after a deadline, I walked unknown streets and streets I know too well, I got lost in a new neighbourhood, I climbed up and down the hills of the city. It is important to note that all walks were performed in the city; their space is the urban environment.

Overall, I walked every time I could, keeping the intention of *A walk, a text, a week* in the back of my mind; at least once a week, I tried to reserve about three hours for a full session of writing in movement. Sometimes, texts appeared on the way to dance class or while searching for a mobile phone repair spot, sometimes I walked the entire evening and not a single phrase came out. I observed both, the unexpected creation and the failure.

This brings us to a reflection on the walks themselves: what happens during the spatiotextual dérives? Although each of them is different, it seems that they all have something in common, somehow a pattern begins to appear. Just like in meditation, jam, improvisation, therapy, relationships, sleep, work or learning cycles and the many processes in life, a subtle underscore is present. We can sense it a posteriori and try to map it, although it obviously does not fit all the cases. There are, of course, cases where moments of lucidity open up and writing happens as a flash, as an insight, just like when we experience an instant, momentary connection with another human being, out of a sudden; it may last, or it may vanish.

Where the underscore is present, two movements can be detected (or a circle consisting of two parts). The first is directed inwards and the second is directed outwards. Before I start walking in the city, I belong to a condition of

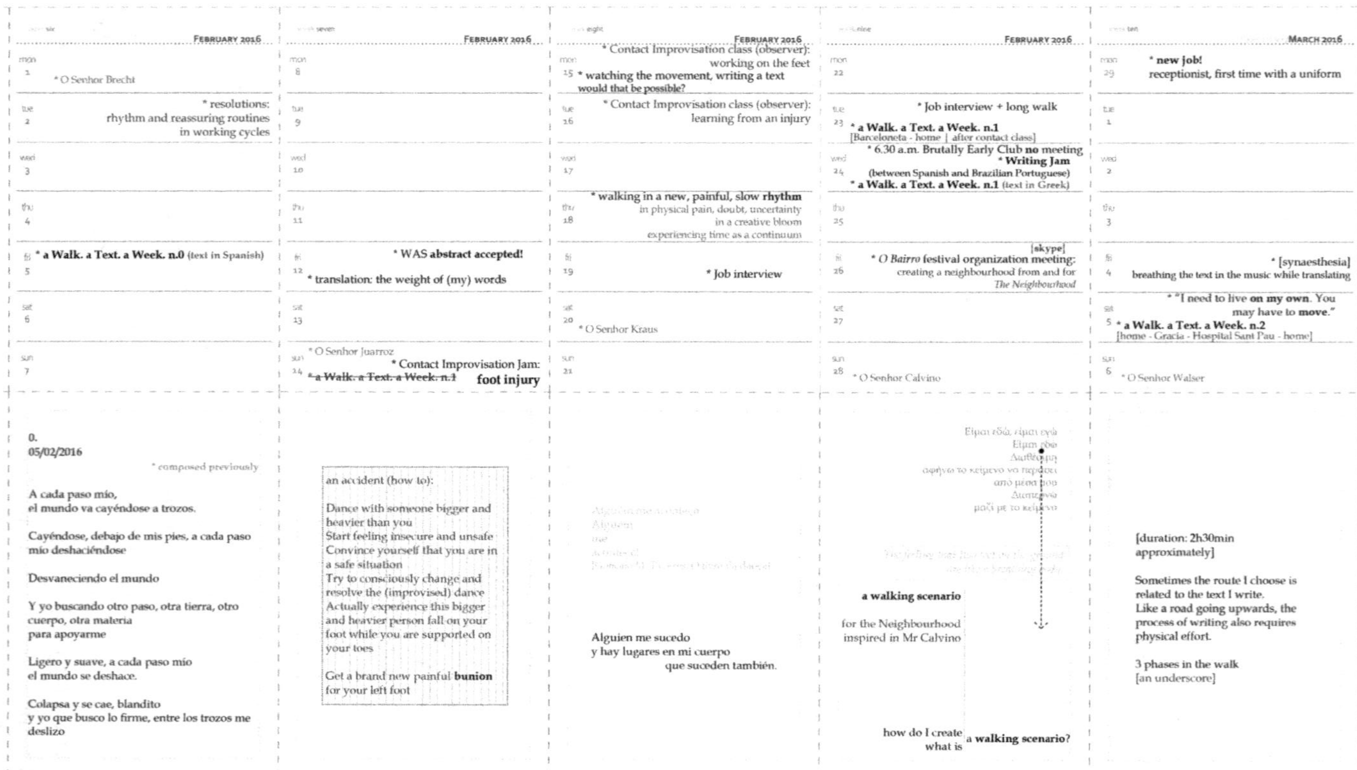

IMAGE 11.4 'A map of twenty-eight weeks' (detail: weeks six to ten)

availability. I am (being) in the world, in the city, and I am available to it; I have a job, things to do, social commitments, friends, family, interests; my phone is ringing and I check my email. When I walk, I gradually absorb the world and abandon this condition of availability; I become present to myself, not available to nor occupied by the world. Then there is a point where something happens, the intensity is multiplied and reaches some peak; after that, I gradually return to the world again and become absorbed by it.

More specifically, when entering a walk, there is this first stage of warming up. I realize and state to myself: I am here. I listen to the things that are happening to me, observe the spines in my body, mind and soul, observe the things that require my attention. As I open my attention towards the city, the sun, the rain, the light, the clouds, the balconies, the trees, the pavements, the people, the dialogues of the people, the smells, the colours, the sounds of the cars, the birds, the bars, the music, the tourists, the sea, the children playing, the dogs (how rich, the city!), as I open my attention, all the worlds and processes inside me put themselves in order. An empty space is created, an inhabitable space.

Gradually, the city becomes a background and I continue walking in my empty space. My footsteps are accelerated and they might start to pronounce themselves. I listen to them. I listen to the words whispered or voiced inside me. Words, like footsteps, one after the other. They sometimes repeat themselves. They are poems, or essays, or letters to somebody. I might need to pronounce them, to pronounce the words whispered to me. I sometimes worry that I might miss them later. I sometimes do miss them. Sometimes I don't. I feel no pain, exhaustion, cold, or hunger; I am inside my footsteps, pronouncing them into words, listening to their music, watching the words dance to it. Everything is footsteps, one foot in front of the other, one word behind the other.

Gradually again, the city recalls my attention. I realize where I am, how far I've come, the sun, the moon, it's hot, I'm tired or thirsty, my feet are burning, my heartbeat is accelerated. That's the point where I realize that I am starting to exit the walk. In a state of fatigue, I observe the city almost indifferently, I observe it as if it weren't for me. The worlds and processes inside me wake up again, somewhat lazy, those too, I observe them as if they weren't for me. I might start smiling at the small insignificant details of the city, I let them surprise me, until I finally return to the world. I am hungry and my feet are heavy. It takes some time until I am fully ready to become available to the world again. A full circle of a walk like that lasts about three hours.

What happens in a writing walk introduces the first layers of writing. After writing in movement, I write. In-movement. With pen on paper, or directly on the computer, on the same day, or many days later, with a sketch or an initial note or with nothing at all, I write. Sometimes the words pronounced in the walk repeat themselves and are pronounced and written again. Sometimes not. Sometimes I create a rhythmic cadence that reminds me of my footsteps. Sometimes I know that a certain word stands for the little stone inside my shoe, or for the dog that barked. Sometimes when I walk past a corner I recognize a sentence I wrote. Sometimes I write and I feel that my words are guided by the sunlight that blinded me while climbing up the hill. Sometimes the words are urgent, I strongly feel I have to write. Sometimes they are patient. Sometimes they are painful but necessary. Sometimes they are in denial, they refuse to be caught. Sometimes they flow. More layers are added with editing.

In the end of this project, I cannot help but wander again. Wonder again. How many layers in space and in time does writing have? How many traces of mental states and how many footsteps are there in a text? How many places has it been before?

Bibliography

Auster, Paul. *Disappearances*. New York: Overlook Press, 1988.

Bakhtin, Mikhail. *The Dialogic Imagination*. Translated and edited by Michael Holquist. Austin: University of Texas Press, 1981.

Bergson, Henri. *Matière et memoire*. Paris: Les Presses universitaires de France, 1965.

Buber, Martin. *I and Thou*. Translated by Walter Kaufmann. New York: Simon and Schuster, 1996.

Carreri, Francesco. *Walkscapes: el andar como práctica estética.* Translated by Maurici Pla. Barcelona: Gustavo Gili, 2013.

De, Certeau, Michel. *The practice of everyday life.* Berkeley: University of California Press, 1984.

Jakobson, Roman. *Six Lectures on Sound and Meaning*. Massachusetts: MIT Press, 1978.

Merleau-Ponty, Maurice. *Phénoménologie de la perception.* Paris: Gallimard, 1976.

Ong, Walter J. *Orality and Literacy: The Technologizing of the Word.* New York: Routledge, 2002.

Perec, Georges. *Espèces d'espaces*. París: Galilée, 2000.

Tavares, Gonçalo M. *Atlas do corpo e da imaginação*. Alfragide: Editorial Caminho, 2013.

Tavares, Gonçalo M. *O Senhor Breton.* Alfragide: Editorial Caminho, 2008.

Tavares, Gonçalo M. *O Senhor Calvino.* Alfragide: Editorial Caminho, 2005.

Tavares, Gonçalo M. *O Senhor Walser.* Alfragide: Editorial Caminho, 2006.

Tavares, Gonçalo M. *Libro de la danza.* Bilingual edition. Translated by Aníbal Cristobo. Barcelona: Kriller71ediciones, 2015.

Walser, Robert. *The Walk.* Translated by Susan Bernofsky and Christopher Middleton. New York: New Directions, 2012.

Weil, Simone. *An Anthology.* London: Penguin Books, 2005.

CHAPTER 12

At the Crossroads: Writing Spaces between Academia and Embodiment

Shelley Smith

Abstract

Writing in the fields of architecture and urban design has become more prevalent. Now seen as a professional tool alongside drawing, writing is used to both clarify spatial ideas and communicate them, and as the building arts have moved into the realm of research, writing has become more 'academic'. In the practice of architecture, a cartographic position is adopted in order to gain an overview, to structure, but it is a removed view devoid of detail and spatiality. The 'academisation' of architectural writing can be viewed as a cartographic positioning, raising the question as to whether it has then become distanced from the very object of its inquiry – space? In the Autumn of 2015 I travelled to a small village in Mexico intending to analyse research data collected elsewhere and to write an academic paper from the quiet and removed reserve of this neutral location. Instead, my emotional response to the space I found myself in triggered a radically different kind of writing – of *this* space. I found that the intentions of my neutrality were challenged by this space, and that a more sense-based approach and sensual embodied spatiality came to the forefront. And there I stood literally at the crossroads. This chapter explores the space of the crossroads of my own writing practice by setting up a dialogue between the reflexive, cognitive voice of academic writing – a cartographic position – and the sensorial, perceptual voice of non-representational inquiry – a more poetic, embodied practice. The dialogue is facilitated by interweaving theoretical positions with 3 multi-media narratives made in Mexico: *Walking Luna*, *Perspective* and *258*. In this way the form of the chapter itself explores the interaction between different ways of writing spaces and the spaces of this writing.

Keywords

Space – spatiality – embodiment – academic writing – communication site analysis – perception – architecture – urban design – non-representational theory

•••

 | DOI:10.1163/9789004394315_013

1 The Ticket

I am an architect, an urbanist, I work with cities and space. In the Fall of 2015, after having completed a period of research in Detroit, I travelled away to Ajijic, a small village in Mexico. The intention was to hole myself away in a hermetically sealed, quiet and removed bubble of space in order to make 'real' in academic written form, the data I had collected via interviews made, cases registered and sites examined elsewhere. Instead, I found myself in Mexico completely and utterly taken aback with the place and space I found myself in. So much so, that I packed away the original object of my research and began writing about *this* place and its spaces. What this place brought forth in me, and how it poured from me in a style so different than the academic style I had painstakingly cultivated over the years surprised me – jolted me into another state – created a new form of contact on many levels but with words as the crafting medium. I found that I could not access this space from a cartographic position, above and removed, distanced and outside. Accessing the space that had opened up to me required something other than the cartographic. It required the commitment of an embodied and sensual response. And truth be told, I seemed powerless to do anything but that. And there I stood literally *in* the crossroads, in a space of transition – having arrived from a distant location only to find that I had missed the connecting flight. But a new journey had already begun, and I was being propelled forward on a path into new spaces of writing. And that, in the words of the poet Robert Frost, from his poem *The Road Not Taken*, 'that has made all the difference'.[1]

This chapter intends to explore my position at the crossroads – a pause in the movement from an academic writing practice to writing as an expressive, sensorial practice. A practice that draws from an analytical research background as an architect and urbanist, but that following the heart, attempts to express a sense of experienced space using writing to reveal this – and share it. This chapter then is a story, it is a journey and it is an experiment. It delves and divulges – it tries to understand and to communicate. This happens within space, leads to a space and is about space. A threefold spatial journey addressing experience and expression.

The crossroads in itself is also a spatial concept. It is a place arrived at – a space in the horizontal plane signifying a stopping, a pause, a coming from and a going to and as such it is a transitional space, a liminal space, a threshold. But nevertheless, a space in its own right. Within the space of the crossroads

1 Robert Frost, 'The Road Less Taken', *Mountain Interval* (New York: Henry Holt and Company, 1916).

3 stories were crafted that have a more performative quality. This chapter takes its point of departure in a personal realisation that my way of writing and role that it plays in my profession has not only changed, but that it has in the process changed my own role as a writer of spaces.

2 Departing

Here at the onset of the journey, a clarification should be made, and the suitcase needs to be packed. Space although heralded as one of the keystone concepts of architecture and urbanism needs specification and examination and with it, the tools of its understanding and manipulation. The object that these musings have been revolving around – space – upon closer examination is a rather confusing concept. Space pops up in a myriad of disciplines as a part of the conceptual framework – in cultural and critical geographies, sociology, philosophy – among others. However, nowhere, but in the disciplines of architecture and urbanism, does space denote a specific area of practice. Here space is the material to be moulded and worked. When we explore the working of space though – we find that the concept that seemed so all-defining and clear is not. In his exploration of the architectural vocabulary, historian Adrian Forty indicates,

> Much of the ambiguity of the term 'space' in modern architectural use comes from a willingness to confuse it with a general philosophical category of 'space'. To put this issue slightly differently, as well as being a physical poperty of dimension or extent, 'space' is also a property of the mind, part of the apparatus through which we perceive the world. It is thus simultaneously a thing within the work that architects can manipulate, and a mental construct through which the mind knows the world, and thus entirely outside the realm of architectural practice (although it may affect the way in which the result are perceived).[2]

And this somehow by raising ambiguity, makes things clearer. Gives a kind of 'aha experience' of how space – seemingly understood and straightforward – creates both a physical framework and provides the perceptual sensorial capacity that contributes to how we are in the world as bodies in space and of space – in the measure of material and the materiality of space itself. It brings the ordering of space and its inherent size and extent to the forefront

2 Adrian Forty, *Words and Buildings* (New York: Thames and Hudson, 2000), 256.

of practice. However, the sensorial capacity is attributed to philosophical categorisation and placed *outside* of the realm of architectural and urban practice. The body and senses become removed from the concept of architectural space.

Walking Luna

IMAGE 12.1 Walking Luna, 1–3

Luna – descending from a proud race,
but still in need of daily exercise. My tempo dictated to by the nose of a dog.
Outside circumstances.
Walking Luna is not so much walking as stopping.
I stop. I stop. Impatient to be moving on.
Blindly waiting – eyes rolling to the empty sky.
And suddenly with a change of focus,
I see.
Not up there, but underfoot.

IMAGE 12.2 Walking Luna, 4–6

Another world,
rich and slow,
full and tactile.
I move between the stopping. Feeling through my feet.
Being told stories of cultures – and living,
of making and doing,
of moving through space – in time.

IMAGE 12.3 Walking Luna, 7–9

I can lift my gaze and take in more – see more – notice more.
What a gift the nose of a dog is.
Opening up minute spaces for the exploration of landscapes.[3]

3 Packing the Suitcase

The suitcase taken on this journey is packed with tools needed to understand and work with space.

3.1 *Cartography*

When confronting the vast object of space, architects and urbanists need to 'get a handle on it' – quite literally make it an object that relates to the hand – originally through drawing. Spatial representation employs the use of scales to make space manageable – to preserve dimensional relations, and to tame the unruliness of space by 'cutting it down to size'. In order to order space, expanses must be miniaturised, cities must be reduced to a table top or screen. Space is made to fit. Cartographic practices have allowed for the mapping, defining and ultimately the working of enormous areas, given the possibility for overview, and provided an understanding of structures and infrastructures, organisational patterns, typographies, typologies and landscape features. It is a useful tool. This can be seen in relation to the duality inherent in the concept of space as described by Forty. The cartographic view takes place at a distance, indicates 'dimension and extent'. It is a practice of providing overview and favours the visual sense, the removed view. It this flattened, two dimensional picture, space as a perceptual capacity disappears. The body, and thereby the senses, are not present in this view. Although the cartographic view is useful for gaining overview, it negates the possibility of sensorial and sensuous

3 The nine images of this section (Walking Luna) were created in 2015 and are reproduced with permission of the author.

engagement. The sensing body is excluded – it is nowhere to be found here – it has nowhere to be.

Finnish architect and essayist, Juhani Pallasmaa has written extensively on bodily, haptic involvement in the spaces of architecture and urbanism. In his influential book, *The Eyes of the Skin* he writes,

> Touch is the sensory mode which integrates our experiences of the world and of ourselves. Even visual perceptions are fused and integrated into the haptic continuum of the self; my body remembers who I am and how I am located in the world.[4]

The sensorial experience of space through fingertips, 'feeling through my feet', of smelling and tasting, is an embodied engagement of the senses – not of the removed and distanced view. The embodied, perceptual experience gives the potential for sensorial capacities, three-dimensionality, spatiality, detail. It is a change in focus from distance to involvement. It is an insider perspective.

3.2 *Words*

The product of my research is predominantly writing, despite the fact that my training is as an architect and urbanist. There is a specific type of writing that is thought 'acceptable' for research – and the necessity of publication – in the community of academics of which I now belong. Writing has become an important form of expression for me. It is an exercise of thinking and communicating, but it is also a vehicle for making real, for reflection – and I also realise – a creative endeavour. The process of design and the process of writing are similar and here there is also a corollary with the cartographic and the embodied. Academic writing in the fields of architecture and urbanism can be called to task though, particularly when addressing the experience of space. The dry, but thorough, heavy, extensively referenced and strictly argued paper as a means of expression can fall short in terms of the object of space being described and accounted. There are similarities between the 'academisation' of architectural and urban writings and the cartographic approach to working with space. The academic perspective is also a removed gaze – an objective view that negates the presence of the subject. It challenges desire to express the experiential qualities of space and culture and – yes – even passionate responses to places. I have found in my training as an architect and urbanist I have developed a lens through which to view, a critical eye and keen analytical

4 Juhani Pallasmaa, *The Eyes of the Skin* (Hoboken, NJ: Wiley & Sons, 2012[1996]), 12.

skills but it is my heart that now is speaking, forming its voice. I find that my architect's eye now connects to my heart and the communication changes form. The training of the architect frames, the eye of the architect informs, but the heart of the architect now awakened, is seeking a new language – a new form of expression. The object of my profession, space, is now beckoning me to approach – and to approach my writing in a different way – as a way to include sensing and to facilitate the sharing of the experience of writing spaces, and perhaps writing as space.

Anthropologist Tim Ingold is a proponent of regarding the tools of our academic trade in a new way. He says,

> We are suffering, in academic life, from a surfeit of words. It would not be so bad if these words, like good food, were rich in flavour, varied in texture, and lingering in the contemplative feelings they evoke. Carefully selected and well-prepared words are conducive to rumination. They enliven the spirit, which responds in kind. But the fact that word-craft of this kind has been hived off to a restricted domain, known as poetry, is indicative of where the problem lies. If writing had not lost its soul, then what need would we have for poetry? We go there to find what is otherwise lost.[5]

This is a departure then, here at the crossroads – from traditional academic writing – to the expression of embodied spaces.

Perspective

IMAGE 12.4 *Perspective, 1*

5 Tim Ingold, 'Foreword', *Non-Representational Methodologies* (New York: Routledge, 2015), viii.

Sloping cobbled streetscapes.
Paved gated estates.
Living Mexican or living Gringo.

IMAGE 12.5 Perspective, 2 and 3

The difference between being inside and being outside.
I move through both worlds – not part of either, but welcome, for now in both
as a non-committal and temporary feature.
The visitor.
With my architect eye, I gather impressions,

IMAGE 12.6 *Perspective, 4*

take neutral note, practice my removed gaze.
And then, I turn a corner –
and this place assaults me!

IMAGE 12.7 *Perspective, 5*

First with colour.
So much colour that I am overwhelmed.
And then with hundreds,
of years,
of history.
Beside me, above me, now in perspective
The massive flowering wall bordering the gated Gringo community,
extending visually into the mountain.
What came after disappearing into what had always been.
Cultivation dissolved into the indomitable indigenous.
The fortress made small by the mountain.[6]

IMAGE 12.8 *Perspective, 6*

6 The six images of this section (Perspective) were created in 2015 and are reproduced with permission of the author.

4 Looking out the Window

If we are to reconnect with a sensuous and embodied form of contact with space and address this in research and the communication of said research, there needs to be an 'opening' in how academic research is understood, practiced and 'judged'. In *Sense of the City – an alternative approach to urbanism*, editor Mirko Zardini remarks,

> In recent years, the human and social sciences, from anthropology to geography, have undergone a 'sensorial revolution' in which the 'senses' constitute not so much a new field of study as a fundamental shift in the mode and media we employ to observe and define our own [*architecture and urbanism, author emphasis*] fields of study.[7]

Interestingly enough, the disciplines Zardini mentions here are also the disciplines mentioned earlier employing space as a part of their conceptual apparatus. In these fields a movement to evaluate how we do research, the practice, methods and expression is underway through Non-Representational Theory and Method. It activates embodied presence in research, attempts to lessen the distance between the object of study, the studying of it and the way this is communicated. Non-Representational Theory stems from the work of Nigel Thrift and Thrift identified 7 points that describe the direction non-representational research should take.

Very briefly they can be described as follows:

1. A focus on events and 'Capturing the 'onflow'…of everyday life'.[8] Everyday life is sociality and it is perceptual in nature and characterised by movement – a lack of stasis.
2. A favouring of the study of relations and relationality. This employs the use of 'and' instead of 'is'.
3. Activities and practices, actions and performances and bodily involvement in these is important. This is concerned with the becoming that this entails.

---

7 Mirkos Zardini, 'Toward a Sensorial Urbanism', *Sense of the City – an Alternate Approach to Urbanism*, edited by Mirkos Zardini (Montreal; Canadian Centre for Architecture, 2006), 22.

8 Philip Vannini, 'An Introduction', *Non-Representational Methodologies*, ed. Phillip Vannini (New York: Routledge, 2015), 3.

4. Bodily interaction is also with objects and here ANT[9] is activated with the notion of assemblages – humans and objects co-adapting.
5. Non-Representation is intended to be experimental. It is not concerned with control and prediction or representation, but rather invites new ways.
6. Bodies!! The inclusion of bodies is also the inclusion of emotions, passions, feelings.
7. 'Boosting aliveness'[10] The notion that strangeness itself is 'the locus for new forms of neighbourliness and community'[11] and new forms of contact.

Non-Representational Theory provides a refreshing approach that by including the body and the everyday to the extent it does – opens for the inclusion of subjectivity, emotions and other forms of interaction between subjects and objects in the realm of research and it opens for experimentation in academia. In addition, it questions the ways in which we present academic research and the affect this has on reception.

Tim Ingold taking a perhaps an even more 'radical' position regarding academic writing practices states.

Surely, our reflections on ways of working cannot be confined to matters of style and composition. They must also extend to the instruments we use, and their orchestration. How does the keyboard compare with the pen, pencil, and brush? Let's try them out and see. Perhaps, then, we will find that working with words, the writer can once again become a draughtsman or an artist or even a musician of sorts. We might cease our endless writing about performance and become performers ourselves.[12]

This opens the door considerably for 'what is legitimate' in terms of research. It makes it possible for the tools of architecture and urbanism to take flight and to contribute actively to the research, understanding and communication of space, and to experiment and develop ways to impart the sensorial experiences of spatial explorations.

---

9 ANT – Actor Network Theory as affiliated with philosopher, anthropologist and sociologist Bruno Latour. See e.g. Bruno Latour, *Reassembling the Social – An Introduction to Actor-Network-Theory* (Oxford: Oxford University Press, 2005).

10 Phillip Vannini, 'An Introduction', 6.

11 Phillip Vannini, 'An Introduction', 6; E.L. Santner, *On the Pyschotheology of Everyday Life*, 6, quoted in Nigel Thrift, *Non-Representational Theory: Space, Politics, Affect* (London: Routledge, 2008), 14.

12 Tim Ingold, 'Foreword', viii.

5 Arriving

258
I passed 258 on several occasions.
It looked over the lake – perched high up on the mountain's foot.
I looked through it.
Mañana. Mañana.

IMAGE 12.9 258, 1 and 2

It showed me its bones,
waiting for covering,
waiting for hands.
Not begging – but in repose.
Building materials stacked lowly at the point of entrance.
Not a barrier as much as a demarcation of in and out,
of here and there,
of someone else's
- and of my own.

IMAGE 12.10 258, 3 and 4

258.
The marking of a future.
Ready for the day there was enough.
But there is patience here.
Finishing is open to interpretation.
And it isn't that it is not.
It is,
in its' bareness,
its' state of in between,
in its' becoming,
reminding.

IMAGE 12.11 *258, 5*

I passed 258 on several occasions and it did not change.
But I did.

Bibliography

Forty, Adrian. *Words and Buildings.* London: Thames and Hudson, 2000.

Frost, Robert. 'The Road Less Taken'. *Mountain Interval.* New York: Henry Holt and Company, 1916.

Ingold, Tim, 'Foreword'. *Non-Representational Methodologies*, edited by Phillip Vannini, vii–x. New York: Routledge, 2015.

Pallasmaa, Juani. *The Eyes of the Skin.* Hoboken, NJ: Wiley & Sons, 2012[1996].

Vannini, Phillip. 'An Introduction'. *Non-Representational Methodologies*, edited by Phillip Vannini, 1–18. New York: Routledge, 2015.

Vannini, Phillip, 'Non-Representational Theory and Ethnographic Research'. Working Notes, 2015. Accessed 3 June 2016. http://ferrytales.innovativeethnographies.net/sites/default/files/Non-Representational%20Theory%20and%20Ethnographic%20Research.pdf.

Zardini, Mirkos. 'Toward a Sensorial Urbanism'. *Sense of the City – An Alternate Approach to Urbanism,* edited by Mirkos Zardini. Montreal: Canadian Centre for Architecture, 2006.

CHAPTER 13

The Artist as a Writer

Esthir Lemi

Abstract

'The tough and the cute dancer' started in 1998 as a collection of photos while using the technique of chronophotography, depicting black and white images of dancers on the move. The film was analogue printed, and in 2013 the photographic material was re-worked via the photosgravure technique. In 2016, 18 years after the first photos were taken, I digitally scanned and reworked these images into an art book. Ten pictures were selected and have been used as inspiration for nine texts. This work-in project of a visual artist and composer with the lifetime hobby of classical ballet started with a photo-shoot. It has been finalized in the form of a book based on text. This process follows the end of the analogue era in printing via digital technology while at the same time it manifests how an artist can express herself in words or her thoughts will vanish, in the same way memories do if we don't keep a record of them. On this presentation, I focus on the artistic process of cultivating artistic skills in musical composition, design and embodied cognition that end up in common rules in writing. The author started with the safe move of common space in art that is 'definition' and defines nine words significant for her career in the same fashion that nine ètudes do in writing. In this chapter, I present a method I consider as a common ground of all arts for the realisation of these ètudes.

Keywords

Chronophotography – analogue – digital – photogravure – embodied cognition – ètude – composition

• • •

1 Introduction

Can a painter use the methodology that she has learned in her studies in order to find common principles in writing? The art book I have just published is such an endeavour. Since artists are always asked to explain their artworks,

 | DOI:10.1163/9789004394315_014

and some of them are good at both art and mutual analysis,[1] there must be something more like inner structure than talent between the two disciplines: visual art and the art of writing. My presentation is about extending this relationship to two more artistic disciplines: musical composition and dance. As my aim is to communicate the material I explore to a bright audience, I believe language is a skill that an artist has as a tool, no matter what their discipline, in some kind of secret programming[2] close to poetry. So even if we are not writers, we communicate in words. My point is not that any artist has the skills to write a novel or to create fictional characters and plot. However, words are there for artists to create their personal vocabulary, hence to create definitions. I call this exercise 'ètudes' and I consider some similar methodology in all arts. Therefore, the title of my book: 'the tough and the cute dancer' refers more to the inner structure (methodology) —TOUGH— and the aesthetic flow —CUTE— that creates a pattern able to adapt itself and develop in time in any material possible. Most important is the communication skill of a secret language, which is probably always based on memory and the way we define images, words and feelings, as well as on our capability to share them in public. Therefore art forms a communicative system, based on our senses. As 'ètudes' I define small texts in a three paragaraph length. Each text describes in an open form a definition given as a title.

2 Captioning

In 1998, as a student at the School of Fine Arts in Athens, I started working on a series of chronophotographies of dancers from the Metsis Studio ballet school, where I used to spend my afternoons as a hobbyist dancer. I set up a darkroom in my home kitchen and started reworking these images for a couple of years. I kept it in a messy archive that I came to redefine here and there during the years in my free time. In 2013, I chose a selection of this portfolio and re-worked it in a heliogravure (printing) technique. In 2015, I selected ten images from the whole work and reworked them digitally. It is interesting how my hesitation to consider this work as finished during the last decade made me follow different techniques from analogue to digital and thus to have a record of 'a threshold'

1 David Sylvester, *The Brutality of Fact: Interviews with Francis Bacon* (NY: Thames and Hudson, 1987); Calvin Tomkins, *Marcel Duchamp: The Afternoon Interviews* (NY: Badlands Unlimited, 2013); Jack Flam, *Matisse on Art* Revised edition Documents of Twentieth-Century Art (L.A.: University of California Press, 1995).

2 Abraham Moles and Joel E. Cohen, *Information Theory and Esthetic Perception* (Chicago: University of Illinois, 1969).

IMAGE 13.1 'À terre'

to a new age, since all of these processes are now considered as old-fashioned; it is really hard to access the material and print it, as it requires vintage equipment and out-of-stock supplies.

When I came to my final selection I had to choose titles. I see this as the moment when the writer inside the artist grows: he/she has to define in words, not an explanation to the picture but more, I think, a word that will function as a link to communication with the audience. I selected nine words that I wanted to define for years. These nine words followed me (I could also say that they guided me) during my artistic life as a professional in choosing directions in ethics. I call this 'my private vocabulary'. These words are in German, English, Greek and Swedish.

1 Abendfüllend
2 The tough and the cute dancer
3 Παράγωνο (oblique)
4 Weakness
5 Baglæns
6 Execution
7 Numb
8 Å
9 Wir lassen uns überraschen

The order in which I put them in my book is connected with morphology in music and the way I learned to use space and time during my studies. Kandinsky observed this common space in music and art in the beginning of 20th

IMAGE 13.2 'Dos au public'

century[3] so the code is well known. I will try to explain how a composer organises development in a musical piece in a way that is similar to creating chapters.

3 The Art of Editing and Musical Structure: from Heliogravüre to Digital Scaling Techniques

We know from program music from the romantic era that a musical idea can reference the title. The connection between this idea and the whole structure through the process is realized via melodic form that is communicated in the course of the composition; the audience follows the thread in transformation and restatement in variations and repetitions. If a title is seen as a definition that will be explained in no more than three paragraphs, then we have an exercise in writing that cannot fail. This process has a similar 'abstraction' to music, i.e. to patterns in the morphology of music. When defining the title for three paragraphs and there is no need to narrate a story, any memories and a personal perspective in random order can be easily organised, like drops of oil falling into a glass of water. Due to the three-paragraph limit and the title

3 Jelena Hahl-Koch, *Wassily Kandinsky und Arnold Schönberg. Die Briefwechsel* (Munich: Taschenbuch, 1999).

that consists of only one word, it is impossible to stray from the subject (as it is almost impossible to lose the melodic line in the very first minute). We have cultivated such repetitive models through myth from the beginning of human culture. Claude Lévi-Strauss analised this phenomenon throroughly. Contemporary theorists in musical perception consider this process of anticipation and repetition as the inner structure in memory where we artists actually project and develop our material. The audiences' anticipation and the limits in human perception form the rules for the development of the material. In other words, it seems that in every new artistic work we follow the same path, a reworked structure that has been processed through the years, and usually we add the least possible amount of information that functions as the individual signature. The pre-structure is already there, and it is 99% of the work. In my book and first writing endeavour, I decided that the material I had was going to be developed in the form of ètudes. In this way I already had an old recipe, a successful structure to follow which also contains all 'magnets and attractors' that keep the outcome well-formatted around one strong 'agent' which is the title; the definition; the theme.

4 Lost in Translation: Revealing the Hidden Negation in Meaning?

The process of the nine ètudes consists of a theme that is then expanded upon in three paragraphs. The three paragraph-scheme is the familiar ABA rondo form:

A: Beginning/ introduction
B: Exposure
A: closure

Since the time of the prelude is short, the brain remembers and creates a connective meaning on its own. This is how composers use melody, and even if we are not able to follow the melody throughout the whole process, it is still hidden in background. This is also why musical forms such as the *sonata*, the *lied* form that we use in songs for centuries, and the *fugue* have had such successful and broad use. Our brain is constructed in this way in order to follow a piece of information for a short while. In other words, this three-paragraph model of the etudes has a strict structure for a successful improvisation.

I originally wrote all nine texts in Greek, as it is my native language, following the above pattern. When I came to translate all the texts, I collaborated with professional linguist in both languages: Greek and English. Interestingly, the English translation not only affected a new perception of the text but also made me look back to the Greek version in order to define and measure the weight of each word in its meaning in the original text. One of the most thorough observations was about the use of negation in contemporary Greek language for a probable reason of optimization in the fact that we are idle when we want to define something, using antonyms that in our head form the opposite of a confirmation, therefore a negation:

> χρόνος άχρονος (time-timeless)
> πιστευτός απίστευτος (credible-incredible)
> βάσιμος αβάσιμος (founded-unfounded)

In the case of the Greek language we seem to erase from our memory some words (usually of ancient Greek derivation, not given to antonyms) because we find it easier to use one word and its antonym (with a negative prefix) instead of two different ones, without bothering to look for a word which would mean, say, 'non-time' or 'non-believable'.[4]

Having realised this inner function of language, with the underlying measure of glumness and entropy it entails, I decided that the first exercise in the ètudes would be to try and find different words for each concept instead of resorting to the convenience of opposite pairs. As far as possible, these words would not be each other's negative — and if they were, this would be deliberate. This is also a psychological way to defrag one's memory from negative forms and remove any secondary dramatic elements.

For all the cliché of the phrase "a picture is worth a thousand words", I find that when we see an image and recall its description, this entropy is not as powerful as when we translate from one language to another. A picture is worth a thousand positive words. In the case of program music (which expresses an image), the powerful dynamic of emotions functions like a game of such intelligence that it cannot act depressively even if it describes horrors. I believe this is because music, as a structural mechanism and through modulation, avoids entropy by its very nature. Translating from image to text or from image to music contains a gesture that reflects a holistic experience, and this has common mechanisms with the musical mechanisms of modulation. The widespread use of opposites in contemporary Greek certainly reflects the current crisis but also the changes imposed by the needs of communication and simplification and the mixture with other cultures.

4 Απόστολος, Βενατσές. *Σύγχρονες Ερμηνευτικές Προσεγγίσεις στη Λογοτεχνία.* (Athens: Εκδόσεις Επικαιρότητα, 1994).

IMAGE 13.3 'Jetè en tournant en avant grand'

5 Creating a Digital Order According to the Performative Process of Analogue Time

The ètude technique has an ABA structure. At the same time, it is accompanied by a picture that shares the same title. The pictures are the origins of these texts and titles, as both were created to describe, explain, and form a unity with them. As soon as I presented the whole project to the editor and her team, a new organization of the structure began; a digital one. The shape of the book also required a re-organization of the material. The team's decision was that the entitled 'Backwards' text, which came with two pictures whereas all others had one, should go in the middle. In this way, in terms of macrostructure the 5th text mirrored the first half of the book against the other half: The process of reading the book is like climbing up from etude 1 to 4, reaching the "peak" at etude 5 and unfolding 2 pictures and then going downwards in order to read the other four. The theme of the first four is performance situations—preparation and on-stage. After the 5th etude there is a slight change of subject and the focus is on the post-performance state of an artist. Each page is a diptych: first you read the text, then you unfold it to see the picture and then you fold it back on itself to turn the page and reach the next text.

On the 5th ètude the reader has two pictures to unfold. One is the analogue picture and the other is the same image printed in heliogravure and mirroring the analogue one. In music morphology this layout is known as 'retrogade' or a 'cancrizans' path (stemming probably from poetry), and is what gives this art-book its symmetry.

During the illustration process another decision was made which was crucial for demonstrating the sixteen-year process of reworking the same material and its laboratory processing, and pointing to the tactile information of the originals. The cover looks like the back of the prints (scanned and re-worked), so that the closed book suggests much about the contents and materiality of what lies inside; this book is an object that opens and invites one to see the other side in four stages:

1. Upon approaching the covers, which embody the back of the original tactile prints, the reader gets an inkling of the performative work carried out in the laboratory.
2. The reader unfolds the diptych-like pages in order to connect the text with the picture. This creates a pause, a semicolon (;) in time, giving time to the reader to assimilate the text before the picture is revealed at the reader's own pace.
3. The central text comes as a peak pont as a position, a context and a title (with the thematics of performance preparation and performance stage divided in a 4:1:4 pattern); after this peak, a cadence follows.
4. The final text functions as a closure.

6 Epilogue

In this chapter, I presented the process of writing my first artist book. I proposed a technique I called "etudes" for writing small texts constructed in three parts based on the context of defining the title. I have chosen eight one-word titles and one phrase as nine definitions that have followed me throughout my career. I tried to explain these titles according to my experiences on stage. Moreover, I based this process on ten images I took while working on the subject of chronophotography and corporeal movement. I worked on this material using extreme handcraft techniques that are seen as vintage these days, so I consider the form of the book as a manifesto of this process; a last survivor of its era. I have chosen to represent this project in the digital age as a work in process of a book, as I believe that through this nine-etude approach I preserve the ideal of the holistic model in art.

Acknowledgements

The publication of this chapter has been made with the financial support of the Scholars Association of A.S. Onassis Foundation.

Bibliography

Bäurele, Reiner. *Meaning, Use and Interpretation of Language.* Berlin: Walter de Gruyter, 1983.

Brun, Siglind. *Sonic Transformations of Literary Texts: From Program Music to Musical Ekphrasis,* NY: Pendagon Press, 2008.

Jones, Steven. *Antonymy: A Corpus Based Perspective.* Oxford: Routledge, 2002.

Khalfa, Jean. *The Dialogue between Painting and Poetry: Livres d'Artistes* 1874–1999. Cambridge: The Fitzwilliam Museum, 2001.

Kivy, Peter. *The Fine Art of Repetition. Essays in the Philosophy of Music.* Cambridge: Cambridge University Press, 1993.

Lemi, Esthir. *The Tough and the Cute Dancer.* Athens: Cube Art Editions, 2016.

Levi-Strauss, Claude. *The Raw and the Cooked. Mythologiques,* Volume 1. Chicago: The University of the Chicago Press, 1969.

Nagrin, Daniel. *Dance and the Specific Image: Improvisation.* Pittsburgh: University of Pittsburgh Press, 1994.

Snodgrass, Mary Ellen. *The Encyclopedia of World Ballet.* NY: Rowman and Littlefield, 2015.

Snyder, Bob. *Music and Memory: An Introduction.* Cambridge MA: MIT Press, 2000.

Verhoogt, Robert. *Art in Reproduction: Nineteenth-Century Prints after Lawrence Alma-tadema, Joseph Israels and Ary Scheffer.* Amsterdam: Amsterdam University Press, 2007.

CHAPTER 14

Exploring the Written Wor(l)d: Writing As a Spatial Practice

Nathan James Crane

Abstract

The practice of writing is inherently spatial, imbuing it with the capacity to 'think' toward other modalities such as architecture that engage with space and more broadly, the concept of spatiality itself. In this sense, writing (at its core) is interdisciplinary in nature. While there has been a considerable amount of speculation on writing practices which can be considered 'spatial' within themselves, the insights uncovered and the implications for disciplines outside of writing/language remain relatively unexplored in some fields of study, in particular, the author's area of research - interior architecture and design. The chapter will examine, as a condition of its aims, the often assumed self-closing nature of writing and will re-present these opportunities as endless, productive sites of investigation for writers and designers alike. Questions arise therefore regarding the value of these writing practices across disciplines including visual art, literature and architecture, asking us to identify the interdisciplinary value writing can offer to diverse areas of research and practice in the broader context of contemporary spatial thought. This chapter suggests writing can offer possibilities/propensities to diverse areas of spatial practice. It will explore the significance particular forms of writing (such as that of French Postructuralist writers Hélène Cixous and Jacques Derrida) offer to spatial practices through an introduction to the author's current doctoral work on the subject. Finally, the chapter aims to develop and then elaborate on the concept of writing as an 'expanded spatial practice'. This work sits simultaneously inside and outside of itself (again spatialised) as does any writing that engages an investigation of textuality. It is possible, in this sense, to constantly observe and question the role of writing, creating the opportunity to offer an expansive articulation of that which is spatial and contribute alternative perspectives on writing's intriguing complexity.

Keywords

Writing – design – space – Derrida – Cixous – expanded field – interdisciplinary – spatiality

 | DOI:10.1163/9789004394315_015

•••

1 Introduction

> He sat at his computer, typing over and over again, the same line. One word forward, and then two backward. Deleting, editing, layering over and over the same words in a different order, all striving, reaching for that myth: a seamless phrase that will speak everything he wishes to say. Over and over, the digital palimpsest of his computer, iterations of a sentence that returns to write itself over, again. He sat at his desk, writing away at this thing, at this delightfully arduous monster, and followed behind it, as a child might, and wanted for it to speak to him again. To whisper the words of the next line, the next phrase that draws it breath.
>
> And there, on that page, he lived out his life: of idea (l) chasing.[1]

The practice of writing is inherently spatial, imbuing it with the capacity to 'think' toward other modalities such as architecture that engage with space and more broadly, the concept of spatiality itself. In this sense, writing (at its core) is interdisciplinary in nature. Here, in this chapter, I am briefly dealing with concepts within my doctoral thesis, specifically, the idea of writing as a spatial practice.

While there has been a considerable amount of speculation on writing practices which can be considered 'spatial' within themselves, the insights uncovered and the implications for disciplines outside of writing/language remain relatively unexamined in some fields of study, in particular, my area of research - interior architecture and design and its agency for studio teaching.

I completed a Bachelor of Interior Architecture with Honours at the University of South Australia in 2012, and enrolled into a PhD program the following year. During my undergraduate experience I had witnessed a distrust of theory amongst my peers and the profession. As someone who considered themselves a writer first and foremost, I was compelled to examine the practice of writing as a generative device, and moreover, it's potential to expand upon or even create new ways of envisioning my speculative spatial practice.

This chapter will examine the often assumed self-closing nature of writing and will re-present these as opportunities for endless, productive sites

1 Nathan James Crane, "Ideal Chasing," (University of South Australia, 2015)., "Situation Symposium and Exhibition Proceedings" (paper presented at the SITUATION, RMIT University, Melbourne, 2014).

of investigation for writers and designers alike. Questions arise therefore regarding the value of these writing practices across disciplines including visual art, literature and architecture, and identify the interdisciplinary value writing can offer to diverse areas of research and practice in the broader context of contemporary spatial thought.

There are particular forms of writing (such as that of French Poststructuralist writers Hélène Cixous and Jacques Derrida) which offer significant insights and allow us to consider and elaborate on the concept of writing as an 'expanded spatial practice'.

It is a certain practice of writing that I am interested in, one that is valuable not just as a medium for the communication of ideas, but rather more critically, as an interstice in which and through which one can generate spatial understanding. It is this generative, space-making capacity (by which I mean, making space for/in thought) which I believe has agency not only for the conception of space, but in a broader sense, the pedagogical comprehension of a spatial design practice.

2 Writing Is Not a Closed Book

In order to speak about the more complex ideas I would like to canvass later in the chapter, it is important to establish the theoretical context in which I am operating, and the way in which I consider writing to be expansive and generative, rather than just prescriptive or communicative.

Writing is a material one can work with. Like any other textile, it can be stitched, cut, threaded, folded. Writing has an additive quality which enables it to be always making more. Writing is a practice at work: a work (crafted) that is 'working' or that is being 'worked', or is always at work (in others). It is a material which is, to more or less degree, a margin-work which engages a margin (to work).[2]

Writing is a practice of situations, or situated-ness that fluxes between interior and exterior. Here, this discussion, this chapter, becomes a writing situation itself, so that again, another interior (writing site) is situated beside the exterior of the 'margin-archive' (the outside: which is an inside, outside). A gathering of situations. A spatial relationship.[3]

2 Ibid.

3 It is a spatial relationship in that, between the work of others (texts/ideas/philosophies), meaning/sense is 'spread out', or spatialised. For instance Derrida's ideas are inside – engaged – in my writing, but remain outside (as an 'other', separate, but related, or relational) to my writing. The relationship between his writing and my own then takes multiple positions, and thus is takes on a spatial characteristic.

In this way, as a gathering, the writing is always inside something else (another piece of writing, a footnote, a margin), engaged in the practice of making/unmaking an interior situation of some kind. The interior of the writing itself, and moreover the interior of the practice of the writing, is a 'situated event'. It is writing (perhaps, writing-situation/writing-situation-event; here again there is a gathering of words) that writes or is written (as writing); a practice (and a space) which engages the interior/exterior relationship of making and unmaking through the use of mechanisms at play in the production of the text, as a practice-process.

That is to say that my writings engage the very process they explore in their production, they are self-evident provocations, which demonstrate through action, the productive mechanisms they seek to articulate. They simulate their original function, an action called autonymy, as Roland Barthes spoke of in his autobiographical text Roland Barthes:

> The enigmatic copy, the interesting one, is the dislocated copy: at the same time it reproduces and reverses: it can reproduce only by reversing, it disturbs the infinite sequence of replicas ... A thousand examples of this *reverberation*, which is always fascinating: the barber getting a haircut, shoeshine boy (in Morocco) having his shoes shined, a cook making herself dinner ... All of which is *autonymy*: that disturbing (comical and banal) strabismus of an operation that comes full circle: something like an anagram, an inverted over-printing, a breakdown of levels.[4]

My thesis is an outcome of its own working, its own spatialisation. The 'work' (of my thesis, the writing, the ideas) is at work in this way. Within this function, each chapter surveys one particular, text-based or literature-based philosophical concept, practice, or tactic (for example the use of repetition). These ideas, identified and proposed as holding significant and productive implications for spatial practice, form an investigation which pursues the underpinning belief that writing has the potential to inform (a) spatial process.

I draw constantly from French poststructuralist theory and in particular, its relevance to the writing practices discussed here, including bodies of work from Roland Barthes, Helene Cixous, Gilles Deleuze, Felix Guattari and Pierre Joris. My thesis seeks to map out a field of enquiry relevant to my key proposition that these discursive writing practices have crucial and previously unexplored implications and theoretical agency for contemporary spatial

4 Roland Barthes, *Roland Barthes*, 1st American ed. ed. (New York :: Hill and Wang, 1977), 49.

practice(s). This thinking stems from encounters I have had with the perception of space I have witnessed (in certain practices and theories). It is through the medium of text and a deployment of writing tactics engaged by Poststructuralist thinkers, that my project aims to facilitate a disruption at the point of space's inception in thought; to interrupt the accepted perception of space (and its privilege toward made-form: organised, systematic, resolved) in order to reconstitute veiled notions of the practice and expansive nature of space. It is at the level at which space is conceived that this project operates.

It explores thinkers, writing practices and tactics who or which engage with a practice or practices that can be perceived as 'spatial', in order to further expand upon that which can be considered a spatial practice.

I am interested in how these spatial writings can influence or offer new conceptions of space, and how they may be employed in order to provide alternative ways to teach spatial design studios in universities and initiate thought in professional design practice.

These particular literary or language/writing practices 'think' toward space, and hold significance for a design/art process. The spatialisation (in both process and form) of Poststructuralist writers and artists' work (installations/objects/artefacts) has been identified as the key sites or situations (both of which are always multiple) in which to investigate the interstices between these practices and interior architecture.

Just as each situation is specific (momentary, transitory) so too are the particular theories and nuanced writing practices of thinkers like Barthes, Cixous and Derrida. This chapter considers these practices as small interstitial sites which coalesce to reveal a relationship with ideas of space making and breaking.

I inhabit two positions in my research, first as a writer, second as a designer. This affords me two perspectives on ideas, and a vantage point that enables me to observe the relationship between space and writing as it occurs to me. It is my deep interest in and curiosity for this relationship that drives my work.

In a sense, it is as Alan Bennett writes,

> The writer is double. There is the self who does the writing and there is the self who does the living. And they talk. They argue. Writing is talking to one's self, and I've been doing it all my life.[5]

---

5 Bennett, Alan. *The Lady in the Van: The Complete Edition*. London: Faber & Faber, 2015, unpaginated.

3 The Space(s) of Post-Structuralist Writing

Poststructuralism's value to cultural practices has been much discussed, particularly in the areas of literature, criticism and architecture. What remains unexamined, are the spatial writing practices that emerged during its zenith, and the significance they may hold for other modalities, particularly (in my case) interior architecture, which itself is still a relatively new discipline.

Poststructuralist writers have, as a tactic, engaged writing in order to interrogate concepts. For instance, Derrida and Cixous both deploy writing as a way to chase ideas, a practice which through it's doing, produces space (both on the page, and in thought). They engage the technique of *iteration* in order to envelop themselves in language, and as a consequence, come closer to (revealing) the inescapably multi-faceted, expanded spatial nature of writing.

In fact, Derrida's entire *oeuvre* could be described as one extensive iterative approach. His writing is a macrocosm of the more specific tactics he employs in his individual essays and books. By persistently revisiting ideas in new frameworks, or by readdressing them in different perspectives, Derrida is able to present these ideas in expanded contexts in order to challenge their very nature and the accepted meaning we place upon them.

One tactical way in which Derrida very specifically engages iteration within his writing is through seemingly relentless repetition. This is evident in *Archive Fever* (1996) where he writes:

> Let us not **begin** at the **beginning**, nor even at the **archive**. But rather at the word '**archive**'-and with the **archive** of so familiar a word. ***Arkhe*** we recall, names at once the commencement and the commandment. This name apparently coordinates two **principles** in one: the **principle** according to nature or history, **there where** things commence – physical, historical, or ontological **principle** – but also the **principle** according to the law, **there where** men and gods command, **there where** authority, social **order** are exercised, in this place from which **order** is given-nomological principle.[6]

Derrida continues to return to the (approximate) starting point in his thinking, and repetitively interrogates the meaning of the word or stem-sentence,

6 He continues, 'There, we said, and in this place. How are we to think of there? And this taking place or this having a place, this taking the place one has of the arkhe? We have there two orders of order: sequential and jussive'. Jacques Derrida, *Archive Fever : A Freudian Impression* (Chicago :: University of Chicago Press, 1996), 9.

in this example, 'there where', 'order', 'place' and '*ackhe*'. Each iteration is an attempt to get closer to unravelling the idea/topic, or dislodge the work from its limited or accepted meaning. Derrida's attempts at writing on a subject are not so much for the purpose of gaining clarity over concepts, but more for making them available for use, so as to reveal the complex histories of the concept(s). By antagonising ideas, Derrida disrupts a singular or binary understanding, and succeeds in unhinging precedents. Ultimately he is making space, putting a distance[7] between the words' ontological groundings and their typographical resonance. This is Derrida's deferral of meaning working at its best.[8]

4 Chasing

Through writing, we can chase (our) ideas. We can follow (our) dreams into the depths of their inception and return again, later, to move even deeper into their complex architectures. Writing is a vehicle in or with which we can drive ourselves into those depths and careen ourselves into the expansive dreamscapes beyond. To write is to chase, to incessantly scrape away at the surface of an idea - the veneer of most is thin. It is underneath, in its substratum, the subcutaneous flesh, where the light is yet to reach that the blood of an idea dwells.

Cixous chases her writing, following it like a dream, until it's fragile and surprising non-termination, as she writes:

> We are not having the dream, the dream has us, carries us, and, at a given moment, it drops us, even if the dream is in the author in the way the text is assumed to be ... [w]e follow it, things go at top speed, and we are constantly – what a giddy and delicious sensation! – surprised. In the dream as in the text, we go from one amazement to another ... I am only interested in the texts that escape.[9]

Winnie Ha Mitford describes a visceral experience of writing, the fulfilment of this chasing:

---

7 Here, 'distance' is both a moving out/expanding and a closing in/concentration: one is always moving away from something in order to move toward/close in on something else, something other. There is also a distance (space) that can be felt between the physical and the emotional. One can be in close physical proximity to an object/person and feel completely distant from them.

8 See Jacques Derrida and Peter Engelmann, *Différance* (Samlerens Bogklub, 2002).

9 Helene Cixous, *Three Steps on the Ladder of Writing*, The Wellek Library Lectures at the University of California, Irvine (New York: Columbia University Press, 1993), 98.

> I write to chase an idea – a fleeting moment to approach some kind of authenticity. It is an exasperation to feel alive. Often the source of pursuit is nameless and formless, more visceral than it is intellectual, more about the experience itself than its fulfilment.[10]

Though we write to discover, the idea (l) we are chasing is already, at least in part, known to us, before we begin. Mitford calls this the 'writer's kernel'.[11] We must know what it is we are chasing, in its germinal state, for us to set off in search of it. The author writes what is already known in this sense. Writer Roland Barthes expresses this in another way in his autobiographical work *Barthes on Barthes* (1977) where he wrote,

> … he [speaking of himself] rarely starts from the idea in order to invent an image for it subsequently; he starts from a sensuous object, and then hopes to meet in his work with the possibility of finding an abstraction for it.[12]

In much the same way, my creative writing practice, which I call 'Gestures', works to create another space for thinking through writing (in which I opened this chapter). One might motion 'goodbye' through waving, it is a gesture, a non-verbal sign, a signal toward some-thing, merely alluding, without being the thing itself. It sets the stage, it is the *before-I-leave*, and *I-am-leaving-now*.

A gesture is non-verbal, yet it still performs a 'language'. It corresponds to a lexicon of words/images, which may be repeated/rehearsed/acted (again): the gesture operates within an internalised (an interior) vernacular, a specificity.[13]

My creative writing, these 'gestures', moves sideways out of the main body of research, and constitutes (an-other) discussion, outside of the main thesis (what I call the 'vocalising body'), which is later distilled back into the main

---

10 Mitford continues, '[t]he pursuit is felt as an intensity that arises from the psyche and extends throughout the physical body. I clothe myself in narrative and write myself into the scene. I need to resolve the story, and this robe is going to help'. Winnie Ha Mitford, "Writing on the Verge of Fashion: Through the Writing Body, a Beginning of a Narrative" (RMIT University, 2014), unpaginated.

11 Ibid.

12 Barthes, *Roland Barthes*, 99.

13 I am not sure here, whether it is a 'correspondence' with a (set of) lexicon(s), or if they are part of them. Can gestures sit outside of what Barthes calls the 'image-repertoire', or do they evoke their own? See Barthes *A Lover's Discourse: Fragments* (London: Penguin, 1977).

discussion of the research. This distilling is a contemplation, a sitting with or within the ideas. A practice of making space, or giving space for thinking.

These 'gestures' point (at given intervals) to influences, or discussions, and again signals my underlying appreciation for Barthes' textual structure, what he calls 'a kind of salute given in passing'[14] to those writers, or readings that have given a moment of inspiration, or even to acknowledge a distant memory, which lingers nearby, and has been the catalyst for a 'gesture' in writing.

What the practices described here do, in one way or another, or to more or less a degree, is challenge the notion of writing by displacing or decentering the role of writing as something not just to communicate ideas, but to generate ideas, and by extension, generate a transdisciplinary spatial practice.

5 Writing as a Transdisciplinary Spatial Practice

As Linda Marie Walker suggests: '… [a] space of writing is a place. A possible place in-between, as another "something" comes into being.[15] There are many examples where writing has been described or treated as a liminal space or place. Christophe Pradeau describes Judith Schlanger's 'lettered space': '… we are talking of a place where one thinks, invents and creates an oeuvre …'[16]

This spatial treatment of text is also highlighted by Samuel Beckett, protégé of James Joyce, writing of *Finnegan's Wake*:

> … Here form is content, content *is* form … It is not written at all. It is not to be read – or rather it is not only to be read. It is to be looked at and listened to. His writing is not *about* something; *it is that something itself.*[17]

Beckett does this in his own writing too. *Texts for Nothing* (1958) is a key example, which illustrates a kind of space making/thinking through writing. Beckett's texts in this book have no apparent linear narrative that can be easily followed, or an outcome that is resolved. What Beckett suggests is that writing is not only important for its content, but rather that it is form too that has value.

---

14 Roland Barthes, *A Lover's Discourse: Fragments* (London: Penguin, 1977), 9.

15 Linda Marie Walker, "Writing-Design, or, Love of the Word (Ephemeris) (1)," *unpublished* (2000).

16 Christophe Pradeau and Roxanne Lapidus, "Judith Schlanger: Explorer of Lettered Space," *SubStance* 31, no. 1 (2002): 68.

17 Samuel Beckett, in Jennifer Bloomer, *Architecture and the Text* (New Haven :: Yale University Press, 1993), 7.

Expressed differently, writing's value is not only in what is being said, but more the mode/way in which it is delivered, its shape/silhouette/surface. Roland Barthes wrote that it is not the message, but the 'message's transport' that was the focus of his attention.[18] It is not about one or the other, but their relationship.

Writing moves far beyond its oft-misplaced function as (only) a communication tool of the already-done (as a mode to report things after-the-fact); it is, at the same time, to recognise it as a significantly more active and dynamic platform for idea generation and most critically, an expanded spatial practice; it makes space for thinking, and beyond this, for thinking about space.

Bibliography

Barthes, Roland. *A Lover's Discourse: Fragments*. London: Penguin, 1977a.

Barthes, Roland. *Roland Barthes* [in English]. 1st American ed. New York: Hill and Wang, 1977b.

Barthes, Roland. *Sade, Fourier, Loyola*. Translated by Richard Miller. Berkeley: University of California Press, 1989.

Bennett, Alan. *The Lady in the Van: The Complete Edition*. London: Faber & Faber, 2015.

Bloomer, Jennifer. *Architecture and the Text*. New Haven: Yale University Press, 1993.

Cixous, Helene. *Three Steps on the Ladder of Writing*. The Wellek Library Lectures at the University of California, Irvine. New York: Columbia University Press, 1993.

Crane, Nathan James. 'Ideal Chasing'. unpublished, University of South Australia, 2015.

Crane, Nathan James, 'Situation Symposium and Exhibition Proceedings'. Paper presented at the SITUATION, RMIT University, Melbourne, 2014.

Derrida, Jacques. *Archive Fever: A Freudian Impression*. Chicago: University of Chicago Press, 1996.

Derrida, Jacques, and Peter Engelmann. *Différance*. Ditzingen: Reclam Philipp Jun, 2004.

Mitford, Winnie Ha. 'Writing on the Verge of Fashion: Through the Writing Body, a Beginning of a Narrative'. Melbourne: RMIT University, 2014.

Pradeau, Christophe, and Roxanne Lapidus. 'Judith Schlanger: Explorer of Lettered Space'. [In English]. *SubStance* 31.1 (2002): 67–76.

Walker, Linda Marie. 'Writing-Design, or, Love of the Word (Ephemeris) (1)'. Unpublished, 2000.

18 Roland Barthes, *Sade, Fourier, Loyola*, trans. Richard Miller (Berkeley: University of California Press, 1989).

CHAPTER 15

The Literary Spaces of Mervyn Peake's *The Gormenghast Trilogy* Used as a Foundation for Architectural Exploration

Imogen Lesser Woods

Abstract

The spaces of literature have the capacity to provide a rich environment for the imagination. Whilst the spaces inhabited by writers and their readers affect the text, the text also has the capacity to provide inhabitable spaces in which one can spend time. The capacity for inhabitation by the reader is directed by the author and Mervyn Peake invites exploration within his texts. Inherent within Mervyn Peake's *The Gormenghast Trilogy* are complex and detailed spaces created solely through literary language then interpreted by the reader's imagination. Peake provides pauses in the narrative so that these spaces might be explored at leisure. He integrates details that might at first appear superfluous and removed from the narrative, yet it is these moments that provide points from which the imagination can spring and allows his environment to be not only believable but to also form an imaginary reality. Peake creates spaces in which one can linger and experience non-physical phenomena. Using the principles of Gaston Bachelard's phenomenology of poetry, as set out in *The Poetics of Space* (1958), a number of Peake's spaces have been rendered architecturally. A single example of this process, the section of the Library of Gormenghast, forms the basis for this discussion as are the processes involved in drawing this image and the subsequent interactions of this rendition with Peake's text. This chapter takes an aspect of my PhD thesis in order to illustrate how literary places can form the basis for architectural spaces and the key role that imagination and phenomenological understanding play in understanding non-physical environments. The library as a key space in the narrative of the first two novels of the trilogy provides examples of the features of Peake's writing that make it so apt for this research.

Keywords

Gormenghast – Mervyn Peake – Architectural Design – Literary Language – Literary Spaces – Phenomenology

DOI:10.1163/9789004394315_016

•••

1 The Premise of Literary Space as Architecture

The spaces of literature are often complex and imaginative, evoking delight in the reader and physical responses to their atmospheres. The poetic prose that is typical of Mervyn Peake's literary works, allows his descriptions to not only bring enjoyment to his readers, but also evoke a phenomenological sense of spatial understanding. This comprehension of space enables an act of non-physical inhabitation within the places of the text and this understanding of literary inhabitation provides the potential for an architectural examination of these spaces. Gaston Bachelard examined the notion that a poetic text might have the capacity for inhabitation in *The Poetics of Space* (1958) and it is from this springing point that the analysis begins.

The inhabitation of non-physical places is important to the architectural discipline: there is a need to understand environments that have not yet been built, to examine theoretical ideas or virtual components of physical architecture. The placement of one's self, miniaturised, within architectural drawings and models allows for the inhabitation of structures before they are constructed in order to comprehend how they might be experienced. There is also a great deal of verbal and written discussion that takes place during the design process. By using the principles described by Bachelard, specifically that inhabitation does not have to be a physical act, literary spaces can be analysed for their potential as architecture *in potential* (from Latin *potens*, powerful). The analysis of Peake's literary spaces therefore studies the potential that this poetic form of language may be used as a tool for design. In order for these loci (places) to be explored through the conventions of architecture a process of rendering a selected few as sectional drawings and models has taken place and it is this process that is discussed here in connection to a single locus, the Library. The works focused on in this study are; *Titus Groan* (1946), *Gormenghast* (1950), and *Titus Alone* (1959). These books are also known as *The Gormenghast Trilogy* or *The Titus Books*.

2 Mervyn Peake's Literary Language as Inhabitable Space

Mervyn Peake (1911–1968) was an artist, illustrator, poet and writer. His literary work carries the spatial understanding that is inherent in his visual art. So although not architecturally trained, his writing conveys experiential qualities which are both evocative and meaningful. It is this comprehension of space, combined with a considered use of vocabulary, (shown in his manuscripts, held at the British Library) which makes Peake's work ideal for an architectural analysis.

Peake was an instinctual writer, with *The Gormenghast Trilogy* beginning as a personal project. The printed editions of the first two novels text closely follow the chronological order of writing found in the autograph manuscripts. In a radio discussion on *The Reader Takes Over* broadcast on the BBC Pacific Service in 1945, Peake considered his writing methods when putting words in the mouths of his characters:

> One of the things I did was to make drawings of them and check the things they said by the drawings, trying to imagine if that kind of remark could possibly come from that terrible mouth, for instance.[1]

Drawings of the literary inhabitants appear frequently in his manuscripts and these sketches, produced concurrently with the text, form an integral part of his writing process. He took great care to ensure that the words he chose were appropriate and poetic. There are lists in the back pages of many of his manuscript notebooks of words he wanted to include, sometimes with their meanings alongside. He clearly enjoyed the exploration of language as he wrote, allowing those who read his work to partake in his pleasure.

Peake's work features several attributes that make it particularly apt for spatial analysis. The first of these is a tendency to switch perspectives during passages and chapters. This is something that readers are not always comfortable with, but provides the capacity within the text for the narrative to show the reader details of differing scales. Peake does not take a fixed narrative view point; instead he allows the spaces and events to dictate how they are perceived. In the following extract the different points of view he uses to describe the library have been isolated:

> [...] Lord Sepulchrave free at seven in the evening, and sitting in the corner of this library, sunk in a deep reverie.
>
> The room was lit by a chandelier whose light, unable to reach the extremities of the room lit only the spines of those volumes on the central shelves of the long walls. [...]

1 Mervyn Peake, 'Radio Discussion: No. 9 in the Series Entitled *the Reader Takes Over* (1945)'. *Peake Studies: Dedicated to the Life and Work of Mervyn Peake (1911–1968)* ed. G. Peter Winnington, 12.2 (2011), 31.

> In the middle of the room, immediately under the light, stood a long table. It was carved from a single piece of the blackest marble, which reflected upon its surface three of the rarest volumes in his Lordship's collection.
>
> Upon his knees, drawn up together, was balanced a book of his grandfather's essays, but it had remained unopened. [...]
>
> The library appeared to spread outwards from him as from a core. His dejection infected the air about his and diffused its illness upon every side. All things in the long room absorbed his melancholia, the shadowing galleries brooded with slow anguish; the books receding into the deep corners, tier upon tier, seemed each a separate tragic note in a monumental fugue of volumes.[2]

Peake shifts the reader's attention from the occupant of the room, to the space and its qualities of light. He then moves to the specifics of the central table and the books that rest upon it. From there the focus reverts back to Lord Sepulchrave and once again, out from him into the space. This shifting of focus and scale at which details are provided can also be found in his use of time and narrative events. In writing this way Peake often halts the primary plot line, pausing the narrative, in order to illustrate a poetic moment. In this manner the pauses in Peake's text echo the daydreams of Bachelard's *The Poetics of Space*:

> This is one of the many daydreams that take us out of this world into another, and the novelist needed it to transport us into the region beyond the world that is the world of new love. People who are hurried by the affairs of men will not enter there. Indeed the reader of the book that follows the undulations of a great love may be surprised at this interruption through cosmicity. But he only gives the book a linear reading that follows the thread of the human events. For this reader, events do not need a picture. And linear reading deprives us of countless daydreams.
>
> Daydreams of this sort are invitations to verticality, pauses in the narrative during which the reader is invited to dream. They are very pure since they have no use.[3]

Bachelard states that these pauses have no use, and remain pure because of this; they allow the author to explore an idea free of obligation. Peake uses them as key moments of reader inhabitation and places in which the author

2 Mervyn Peake, *The Gormenghast Trilogy* (London: Vintage Books, Random House, 1999), 146.

3 Gaston Bachelard, *The Poetics of Space*, trans. Maria Jolas (Boston: Beacon Press, 1994), 162.

invites us into the space to explore. These daydreams are the sticking points which Robert Louis Stevenson described in a 'A Gossip on Romance' (1882) and the places of heightened dislocation that provides Michel Butor with his displacement when he reads: 'This 'volume', as we say, which I hold in my hand, sets free as a result of my attention evocations which assert themselves, which haunt the place where I am, displacing me'.[4] In becoming stuck within the text, entering it fully, in inhabiting the spaces of the text – not only the spaces of the page, the ink and paper but also the places created by the text – the imagination is able to bring more to the experience than the environment created by the author. In enriching the written spaces, and with the right circumstances, the reader is able to become dislocated from reality and forget the real space they are situated in.

3 An Architectural Examination of Peake's Library

The library of Gormenghast is one of the spaces which Peak returns to in the text a number of times. It contains a series of significant events and provides opportunities for Peake to invite the reader into the space. In his inclusion of descriptions consisting of poetic, atmospheric and sensory qualities along with measured distances Peake forms the very rich and complex environment of *The Gormenghast Trilogy*. This interweaving of measurements with phenomenological qualities is vital to the reader's understanding of the library as a structure and the dramatic events that unfold within it:

> The room was lit by a chandelier whose light, unable to reach the extremities of the room lit only the spines of those volumes on the central shelves of the long walls. A stone gallery ran round the library at about fifteen feet above the floor, and books that lined the walls of the main hall fifteen feet below were continued upon the high shelves of the gallery.[5]

Within this short passage Peake not only uses descriptive qualities of space to show the reader the quality of light, material and atmosphere; he quantifies the height of the gallery that is set within the room and later the size of the single window. There are also many details that can be teased apart by the architectural reader to garner knowledge of the other senses. The reader does

4 Michel Butor, *Inventory: Essays*, trans. Richard Howard (London: Jonathan Cape Ltd, 1970), 32.

5 Peake, *The Gormenghast Trilogy*, 146.

this instinctively as their imagination acts unconsciously to fill in the details. Examples of this inherent understanding of spatial qualities can be found in the materials Peake uses and the contents of the room. In the library the light is provided by a chandelier filled with candles. The quality of light is therefore very particular, as a product of combustion – very different to that of a room with electric lights. The heat and the smells produced as a byproduct of this illumination are integral in the understanding of what a candle is. By describing the chandelier Peake also provides knowledge of the space and its ambiance. The books in the library, leather bound and ancient, add to the layering of understanding: the smells of old leather and paper; the taste of dry pages and dusty shelves; the tactile awareness of the weight and texture of an ancient tome are combined with the physical qualities of the building. It is a stone structure, inhabited by the characters primarily at night; it is dusty and filled with dark corners. There are velvet backed chairs worn through use and a large black marble table that reflects the objects that sit upon it. The reader can become immersed in the descriptions and occupy the architecture alongside the characters.

For the architectural draughtsman, producing images of the space in order to analyse them, these qualities must be examined so that they might be drawn. The draughtsman must exist in a dual state of immersion and objectivity so that they might analyse the data provided by the text but remain aware of the quality of space and its ephemeral components. In creating visual renditions of Peake's spaces it is not only a matter of marking out the dimensions, as he rarely provides enough for a full space, but also inferring qualities. When drafting one can calculate distances from the heights of characters or the length of their stride (provided by Peake) and number of paces they take. It is a process of investigation and iteration which both fixes and alters the spaces that Peake created.

In rendering literary spaces the events of the narrative also provide information about the spaces that contain them. Peake played with time within his novels: the first novel, *Titus Groan*, takes place in a span of less than a year but *Gormenghast* (of comparable written length) takes place in a period of ten years. Peake shows the reader events that occur concurrently but as a bound text these simultaneous happenings are presented in a defined order, so the reader encounters them consecutively. In the architectural section the usual method of dissecting a space with a single slice also defines a fixed moment. The internal volume of the structure is exposed and the events frozen in the instant of the dissection, allowing only a single instant to be revealed.

In order to render the Library of Gormenghast (Image 15.1), with its different states and events, the architectural section was cut again. These additional

IMAGE 15.1 Sectional architectural rendition of the Library of Gormenghast

slices cut perpendicular to the primary section, allow the structure to be seen in different places and different instances to be portrayed. This cranking of the section and juxtapositioning of events forces the structure to be indeterminate in size. Peake did not indicate dimensions for the library as whole, only specific aspects. Thus there are dimensions that remain ambiguous. Peake provides the knowledge that 'The library, though a lower building than these two to which it was joined at either end, was of a far greater length than either'.[6] Its measurements, therefore, are fa Michel Butor, *Inventory: Essays*, trans. Richard Howard (London: Jonathan Cape Ltd, 1970), 32.r from certain with little indication as to how long it might be. In order to render this ambiguity, the cranked section allows for the imagined insertion of additional portions where the vertical lines perforate the image. The structure can therefore be extended; it becomes a building with the potential to be any size beyond the minimum length depicted, with the capacity for an infinite expansion within the imagination. The hidden corners and multitude of books can be extended beyond the boundaries of the page. Peake's library follows in the traditions of labyrinthine literary libraries: impossibly large, yet familiar.

In rendering a version of this space this image dissects not only the form of the building but the narrative events it contains. The section lines allow

6 Ibid., 145.

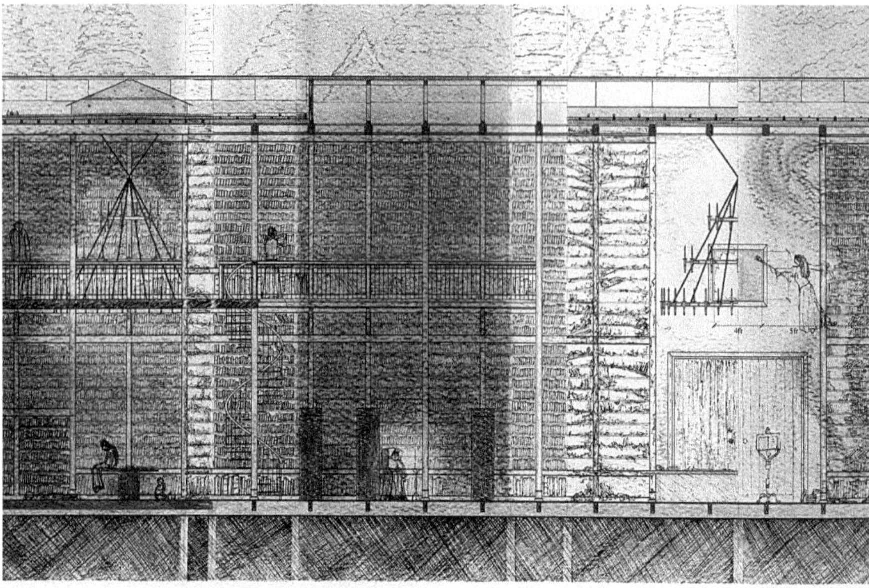

IMAGE 15.2 Detail of the Library showing Sepulchrave reading at night, burned and unburned shelves in daylight and Fuchsia breaking the window during the fire.

occurrences to be placed side-by-side and time also becomes sectioned (Figure 2). The sequencing of events is altered; the nocturnal reading of Lord Sepulchrave is contrasted with the daylight sneaking of the protagonist Steerpike. The day becoming more haunted than the night; hints of the destruction yet to come, co-exist with the library in its pre-fire state and whilst those who have read the plot may be aware of the events that are contained within those who view the space for the first time are not fully enlightened.

4 The Rendition as a Personal, Objective Image

Any rendition is, of course, a personal one. However, the techniques used to describe Peake's space visually involved a process through which it was the text that provided the basis for each decision. Dimensions are set within the drawing so that those Peake specifies are seen. A two stage process was used, beginning with a logical process drawn out through digital lines. These lines, clinical in appearance and drawn from the information provided, remain in the final image. They act as a reminder that the drawing was formed through an analytical process as well as inference and intuition. The second stage was the re-introduction of the atmospheric qualities of the space and the life of the text. Plants, animals and human characters were all added by hand, as were the events of the text. The atmospheric understanding of the space had to be brought out of the digital lines, which enable a false accuracy.

It is easily assumed that a digital image is 'finished' when it is not, just as a typewritten document has a false level of permanence when compared with a manuscript. When one is shown a printed digital image the eye reads a fixed line where it is in reality still uncertain: a return to the digital file allows for alterations. In completing the renditions by hand, the understanding of impermanence is brought back, even though these lines are far more permanent in their nature. There is an uncertainty that can be present in a hand drawn line that becomes too crisp in the digital print. This fluidity of line and expression of purpose that can be read in a hand drawn line is one that allows the eye to imagine through the imperfections. The imagination is stimulated; just as one reads Peake's texts and wishes to wander through the unseen spaces that are hidden behind closed doors, the rendered spaces provide places for the imagination to latch onto. Characters cannot cross the section lines, as this is a step through time, but the viewer can pass through these internal barriers easily. There are figures that appear more than once within the image, co-existing events that the viewer can interpret, inhabit by proxy or interact with within their imagination.

In an architectural section the viewer is shown the space in a predefined way, it is removed from the points-of-view that Peake uses and becomes a stage set within which the immobile drawn figures become merged with the imagination. Yet, through the drawn image of Peake's library, one is able to comprehend in a similar manner to the author. The events which make the space interesting are inherent in its representation, both within the imagination and when rendered. The shifting in scale is no longer present in the section but instead the distortions in time become prominent. The rendition creates a different library to the one that exists within the imagination and is inhabited by readers, but so is any architectural structure. The remembered space is a version of the built form and inhabited space is a personal experience that constantly changes. The use of language to explore these ephemeral qualities allows for an enrichment of architectural design beyond visual renditions.

Bibliography

Bachelard, Gaston. *The Poetics of Space* [(1958) La poétique de l'espace]. Translated by Maria Jolas. Boston, Massachusetts: Beacon Press, 1994.

Butor, Michel. *Inventory: Essays*. Translated by Richard Howard. London: Jonathan Cape Ltd, 1970.

Peake, Mervyn. *The Gormenghast Trilogy*. London: Vintage Books, Random House, 1999.

Peake, Mervyn. 'Titus/Gormenghast Series' Add MS 88931/1/3, British Library, 1939–1962.

Peake, Mervyn. 'Radio Discussion: No. 9 in the Series Entitled *the Reader Takes Over* (1945)'. In *Peake Studies: Dedicated to the Life and Work of Mervyn Peake (1911–1968)*, edited by G. Peter Winnington, 12.2 (2011): 25–41.

Peake, Sebastian, ed. *Mervyn Peake: Boy in Darkness and Other Stories*. Centenary edition. London: Peter Owen Publishers, 2007.

Stevenson, Robert Louis. 'A Gossip on Romance', In The Works of Robert Louis Stevenson, Volume 9, e-book library edition, 134–147. Salt Lake City, Utah: Project Gutenberg, 2009.

Index of Names

Index of Subjects